T0173089

DURABILITY ANALYSIS OF STRUCTURAL COMPOSITE SYSTEMS

LECTURES OF THE SPECIAL CHAIR AIB-VINÇOTTE 1995

Durability analysis of structural composite systems

Reliability, risk analysis and prediction of safe residual integrity

Edited by
ALBERT H. CARDON
Faculty of Engineering (T.W.), Free University Brussels, Belgium

A.A. BALKEMA / ROTTERDAM / BROOKFIELD / 1996

Published by
A.A. Balkema, P.O. Box 1675, 3000 BR Rotterdam, Netherlands (Fax: +31.10.4135947)
A.A. Balkema Publishers, Old Post Road, Brookfield, VT 05036, USA (Fax: 802.276.3837)

ISBN 90 5410 640 9

Contents

The AIB-Vinçotte academic chair

About the chair itself

The AIB-Vinçotte academic chair was set up in 1990, under the auspices of the National Fund for Scientific Research (NFWO-FNRS), to celebrate the centenary of AIB.

This chair lasts for 8 years, and alternates between the various Applied Science Faculties of the Belgian universities.

The chair gives the faculty concerned the opportunity to invite specialists from Belgium and abroad to give lectures on subjects relating to the safety and management of industrial risks. So, in the past, the chair has addressed the following topics:

– In 1991, at Ghent University: safety of welded structures in energy production and energy transport;

– In 1992, at the Université Libre of Brussels (ULB): reliability of industrial installations;

– In 1993, at the Catholic University of Leuven (KUL): damage investigation of materials: basic principles, procedures and techniques;

– In 1994, at the Catholic University of Louvain (UCL): risk management and industrial safety;

– In 1995, at the Vrije Universiteit of Brussels (VUB): durability analysis of structural composite systems.

In 1996, 1997 and 1998, the Polytechnic Faculty in Mons, the Royal Military Academy and the University of Liège will take their turn.

About AIB-Vinçotte

For over a century, AID-Vinçotte has been working to improve safety. In 1872, a young engineer named Richard Vinçotte founded the Association Vinçotte with the aim of improving the safety of steam boilers and steam vessels. In 1890, the Association des Industriels de Belgique – AIB (Association of Belgian Industrialists) was founded with the objective of preventing accidents at work. Similar initiatives were taken around that time in various industrialised countries, and within Europe, these can now be found in the Colloque Européen des Organismes de Contrôle (CEOC – European Colloquium of Testing Bodies), of which AIB and the Association Vinçotte are founder members. Most of these associations came from the private sector, and were later given the task by their respective governments of implementing legally-required inspections to ensure the safety of workers.

To take up the challenges of the European single market after 1992, AIB and the Association Vinçotte decided to merge in 1989, together with Controlatom, a joint subsidiary that was founded in 1965 to carry out inspection and testing with regard to the risks of ionising radiation.

Technological and social change in the 20th century has been an important source of expansion of the range of activities carried out by AIB and the Association Vinçotte, so that nowadays, the various companies in the group cover virtually all the safety aspects relating to industrial risks. In addition to the safety aspect, AIB-Vinçotte has also expanded its activities in environmental protection and testing of the quality and operational reliability of plant and equipment.

This work is all carried out by a total of over 1500 employees in Belgium, and in our establishments in the Netherlands, the Grand Duchy of Luxembourg, the USA, Italy and the United Arab Emirates.

Almost one-quarter of turnover is generated abroad.

Generally speaking, AIB-Vinçotte offers the following services:
– Certification of systems, products and people;
– Inspections and audits;
– Testing and measurements (permanent and mobile laboratories);
– Advice, training, research, expert opinion.

Wherever necessary, AIB-Vinçotte has the necessary authorisations, accreditations and notifications.

The range of industrial applications is very large, and ranges from domestic electrical installations to nuclear power stations, from a lift to the construction of the whole building, from a hoist to a whole factory, from an inspection of equipment to a whole pipeline, etc.

The future is already with us
The future of AIB-Vinçotte is tied up with future of industrial society, which is in turn linked to the changes in society in general and with the requirements that this society imposes on industry.

That means, among other things, higher requirements for safety, environmental protection and operating reliability. At the same time, there is exponential growth in the field of data processing and communications.

AIB-Vinçotte's service provision will have to be continually adapted to the latest requirements of the market. Obviously, it would be better to play a trend-setting role in these changes. To achieve this, we must be able to call on the necessary expertise in house and, if necessary, from external advisors.

Therefore, AIB-Vinçotte attaches the greatest importance to maintaining its excellent relations with the various applied science faculties involved with the organisation of the AIB chair.

This is a special form of cooperation between industry and university which fits in with our vision of a safer, more environment-friendly and reliable industrial society.

Ir Jean Smets
General Manager

Durability analysis of structural composite systems: Why and how?

Composite systems are one of the oldest material and structural answers to resist a specific loading system and history.

A composite system is composed by at least two materials with characteristics on a macroscopic level, and at least one transition surface, or region, between those materials (*). The properties of the composite system are a combination of the properties of the basic materials and the way they interact. The composite has some properties that are better than the simple addition of the properties of the constituents. In most cases we have a continuous phase, the matrix, and a discontinuous phase, the inclusions.

The geometrical form of the inclusions may vary from platelets, for the reinforcement of coatings, over grains and spherical shapes to fibres, from short to long continuous fibres and complete 2 and 3-dimensional textiles. The matrix may be of cementeous nature, a ceramic, a metal or a polymer.

The reason that we call the composite a composite *system* is based on the fact that a composite may be considered as a material but is sometimes an integrated structure, without the possibility to consider separately the material behaviour separated from the structure.

We consider δ as the characteristic length of the inclusions. We have a representative volume element (REV) around a point P, with a characteristic dimension d. For a meaningful homogenisation, we must have the possibility to move the RVE around P in a certain volume of space without change of the obtained properties.

If d^* is the characteristic dimension of this volume we have the homogenisation condition $\delta \ll d < d^*$.

If we consider a structural component, able to transfer forces from their input zone to their output region, this component has generally a characteristic dimension D, who may be a function of the application: tension-compression, buckling and vibrations.

(*)This definition implies that we do not consider molecular or in situ composites, where the property changes are induced on a molecular level.

The use of a material in a structural component needs the condition $d^* \ll D$.

If d^* and D are of the same order of magnitude, the composite system has to be considered as an integrated structure.

Concrete is a material. Reinforced concrete, with long reinforcements, is a structure. In the human body the bone of the skull can be considered as a material, but the thighbone is an integrated structure.

Excellent presentations can be found in two books by J.E. Gordon [1, 2].

The design of structural components with composites does not present great difficulties anymore. Many design codes exist and the menu of available basic materials for matrix and fibres is a very large one. There are no special problems if the concern is about short time behaviour of the component.

The guarantee of a safe long term structural integrity after a complex general, mechanical and environmental, loading history is still a problem.

The different methods developed in order to obtain on a short term basis information on the long term behaviour and the guarantee of a safe residual structural integrity after an imposed life time is what we define as durability analysis.

The presence of inclusions, especially fibres, and even more, long parallel continuous fibres, are interesting for stopping, even only for a limited time, crack propagation. The higher probability of initial defects combined with the presence of residual stresses as result of the manufacturing process are the negative aspects of the composite concept.

For some matrix materials we have to consider the influence of time. Their properties are changing as a function of time, with and without loading. In this last case we have the so called ageing effects. Polymers are viscoelastic and polymer matrix composites will present such time dependent properties.

Environmental factors such as moisture, temperature and radiation have an important influence on some matrix materials, on some fibres and on the interaction level between the fibres and the matrix. Most of those environmental factors are working in interaction with the stress state, e.g. the specific stress assisted moisture diffusion in polymer matrix composites.

During the loading stress transfers occur between the phases in a composite. The level of those transfers is controlled by the characteristics of the interaction region, interface or interphase, between the constituents.

If the composite concept opens the very interesting possibility to produce systems with properties adapted, 'à la carte', to the loading, including the introduction in the system of some specific anisotropy of stiffness and strength behaviour, there is need for a validated methodology in order to perform a good durability analysis.

This book contains a large overview of the actual problems related to the durability analysis of composite systems. After a general introduction to the problem of reliability follows the analysis of the origin of structural failures, the problem of the technical insurance and the legal consequences.

The further chapters are devoted to:

– The application of composites in marine environment;

– The prediction methods for fatigue behaviour;

– The specific time dependent phenomena of some systems;

– The presentation of an integrated prediction program for the residual strength of a composite structural component;

– The importance of the initial design for the structural integrity and durability with emphasis on damage tolerance concepts and the design for non catastrophic failure behaviour. (see also the report of the Committee on Engineering Design Theory and Methodology of the US National Research Council [3]).

Many examples of optimal structures exist in nature. The concept of hierarchical structures in biology is an important guide for future developments of composite systems [4].

In the actual stage of development we are still working with inclusions to be introduced in a matrix followed by a temperature and pressure program in order to manufacture the composite.

The human body, and other biological systems, is not the result of such a process. In the human body all the basic ingredients are present at the beginning and they grow during time in order to arrive at the complete integrated composite system. This is also the basic concept of the in situ or molecular composites where the development on molecular level will result in an integrated composite.

Though those molecular composites will probably be, on the long term, the most interesting path to integrated optimal structural components, today we have available the keys of the macroscopic composite way to efficient structures and even to active or smart structures.

For a larger application of the composite concepts the durability challenge has to receive an answer. With this publication we have tried to give the state of the art in the domain.

This chapter was a tentative answer to the question: Why? The following chapters will develop some answers to the question: How?

REFERENCES

[1]   J.E. Gordon 1968 and 1984. *The new science of strong materials*. Princeton University Press.

[2]   J.E. Gordon 1988. *The science of structures and materials*. Scientific American Library. Freeman & Cy., New York.

[3] National Research Council 1991. *Improving Engineering Design* (Designing for competitive advantage). National Academy Press, Washington D.C.

[4] National Materials Advisory Board, Commission on Engineering and Technical Systems of the National Research Council (NMAB 464) 1994. *Hierarchical Structures in Biology as a Guide for New Materials Technology*. National Academy Press, Washington D.C.

Albert H. Cardon
Professor
Composite Systems and Adhesion Research Group
of the University of Brussels (COSARGUB)

Brussels, October 1995

Reliability, a modern issue in engineering

Marc Van Overmeire

Mechanical Engineering Department, University of Brussels (VUB), Belgium

1 INTRODUCTION

The former US secretary of State for Defense, James R. Schlesinger stated that 'Reliability is, after all, engineering in its most practical form'. [1].

Indeed, in today's engineering practice, reliability demands become more and more stringent, and this is not only true in design, but also in production. When buying or selling products, reliability is playing an important role.

People often hesitate to buy products made of new materials. Their reliability is indeed unknown, and certainly not proven. In this book, the different approaches to validate the reliability of composite materials are discussed. This chapter introduces the important concepts of reliability engineering.

2 RELIABILITY

Everyone is aware of the problems, which can be caused by unreliability: the TV set which is not working at the very moment of an important football match or exciting movie, the automobile which does not start, when having an important meeting.

Manufacturers often suffer high costs of failure under warranty, certainly when a lot of the sold products are defective. Unreliability can provoke high costs to airliners and public utilities. But although everyone is quite aware of the consequences of unreliability, it will be hardjob to find someone who can give a correct and complete definition of the word reliability. For some people, even engineers, it is the same as quality control. Although quality control makes an essentail contribution to the reliability of a product, it does not take the time as parameter into account.

Engineers are trained in deterministic approaches and calculations; reliability is however uncertainty engineering requiring the use of statistical methods. But statistics in reliability engineering are less straightforward than in measurement of human variation such as height, weight or I.Q., as we are dealing with unlikely combinations of load and strength, so-called extreme conditions, and so understanding reliability means a lot of effort, many hours of study and certainly a change of attitude.

Moreover, the amount of useful data is limited. As engineers are currently changing design to improve products, reliability data from any past cannot be used, without adaptation.

After this long discussion, we finally come to the engineering definition of reliability. 'Reliability is the probability that an item will perform a required function without failure under stated conditions for a stated period of time'.

The main running parameter is certainly not in all cases the physical time, it can be another time-related parameter such as mileage, operating cycles, seasonal cycles, number of revolutions e.o. depending on the type of system and its use.

The addition of the words 'under stated conditions' ensures that the item will be used for the function for which it was designed. A 12V DC lamp should not be used on a 220V AC circuit.

Also the correct and detailed description of 'required function' is very important. Does an audio amplifier become unreliable when one of its leds, indicating the delivered sound power, fails? Certainly not, as its function is to amplify the analog audio signal coming from other devices such as radio, CD-player e.o. and the led performs only a (minor important) signalling function.

Of course in our world , the life of a product is limited (as shown in the definition of reliability). The life of the item stops when a failure occurs, but what is a failure and how can we define it?

3 FAILURE AND FAILURE RATE

According to engineers, failure is the termination of the ability of an item to perform a required function.

The way in which the failure appears is called the failure mode.

Failure and fault are not the same: failure is an event while fault is a state. Before introducing the mathematics of reliability, let us first of all try to gain more understanding in the way how systems and components fail.

A component fails if the applied load exceeds the strength. Load and strength should be considered in wide sense for general reliability, but in case of mechanical systems load refers to forces and moments and strength to yield or fatigue strength, strains e.o.[2].

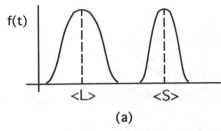 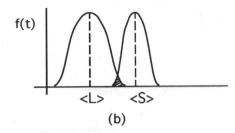

Figure 1. Overlap of strength and load causes failure.

Of course, neither load nor strength are deterministic values, but they are distributed statistically.

Each distribution can be characterised by its statistical parameters i.e. mean value and standard deviation. Failure occurs when the two distributions overlap, as shown in Figure 1. Remark that although the mean value of strength is much larger then the mean value of load – expressed by the safety margin – failure still happens for components at the lower end of the strength distribution, subjected to extreme high loading.

In the life of a component, strength will degrade due to fatigue, wear and corrosion. This will cause an increasing overlap, which finally will lead to the failure of all components.

To describe the evolution of failure as a function of time, the failure rate is used. The failure rate λ is defined as the number of failures per unit time. It is usually given in terms of failures per million hours or per year and can can be presented by Equation (1)

$$\lambda(t) = \frac{N_f}{N_o \Delta t} \tag{1}$$

where N_f is the number of systems, failed during the period $(t, t + \Delta t)$ and N_o the original number of systems.

If the evolution of the failure rate as a function of the relevant parameter e.g. time is plotted, the so-called bath-tub curve is obtained (Fig. 2).

As can be seen, the failure rate varies with the age of the item. Three stages are present in the life of a component or system:
  – Stage A: infant mortality;
  – Stage B: useful life;
  – Stage C: wearout.

The infant mortality of devices, which can be eliminated in a lot of cases by preliminary screening and testing (e.g. burn-in testing for electronic devices), is characterised by a decreasing failure rate. During the stage of useful live, the failure rate is constant (at least for electronic devices), but still depends on the stress and environment conditions. Failures follow a random pattern.

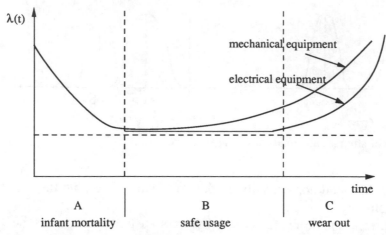

Figure 2. The bath-tub curve.

For mechanical components however, wear, fatigue and corrosion are causing an increase of the failure rate.

In the wearout stage, failure rate increases rapidly and the last survivals will finally fail.

The failure rate which is normally used in reliability calculations corresponds with stage B of the bath-tub curve. However following remarks have to be born in mind.

1. The bath-tub curve published for any given device or item is based on a given set of application conditions, usually a standardized representation of the operating and environmental stresses that the part is expected to sustain during its service life. A change in one of these conditions will displace the curve, relative to the operating time and failure rate axes, and may also significantly alter the shape of the curve or the durations of the three stages represented.

2. The bath-tub curve for a given part will probably differ slightly from one production lot to the next, because of minor variations in tooling, workmanship, fabrication processes, materials, and other manufacturing factors. Consequently, the bath-tub curve is better depicted as a broad band, rather than the single line shown in Figure 2. The 'best-case' boundary represents the optimal combination of manufacturing factors, while the 'worst-case' boundary represents the most adverse combination.

3. Completion, by a production lot, of the infant-mortality stage (burn-in) will not necessarily weed out all failures traceable to manufacturing defects. If remain, they will occur at unpredictable intervals throughout the safe-usage second stage, thereby contributing to the failure for that stage.

4. Every failure occurrence must be presumed to have an explainable cause. Those failures occurring during the second stage that cannot be

readily identified as manufacturing defects are invariably lumped together as so-called 'random' failures because they do not recur frequently enough to justify the cost of identifying the failure mechanism and eliminating it. Indeed, positive identification of certain kinds of failure mechanisms is not feasible because of economic considerations or state-of-the-art limitations in materials availability or fabrication processes.

5. With the possible exception of solid-state devices, all functional parts appear to possess an inherent degradation mechanism in the form of wear-out corrosion, materials aging, or some other physical phenomenon that will cause the part ultimately to fail in normal service, even if the part is never exposed to stresses beyond the levels for which it was designed. That's why the bath-tub curve always has a third stage.

4 EXPONENTIAL RELIABILITY DISTRIBUTION

If we suppose that the component has a constant failure rate during its useful life time, the exponential life distribution will be the result. As neither wear nor fatigue are taken into account, this distribution is not applicable for mechanical systems. However in practice it is frequently used as data bases and books only provide constant failure rates as reliability data.

To derive the exponential model, we define reliability as follows:

$$R(t) = \frac{N_w(t)}{N_o} \tag{2}$$

where: N_w equals the number of systems, still working at time t, N_o the number of systems at time $t = 0$.

From Equation (1) and supposing a constant failure rate gives:

$$\lambda = \frac{-dN}{Ndt} \tag{3}$$

After integration, following expression for the reliability is obtained:

$$R(t) = \frac{N_w(t)}{N_o} = e^{-\lambda t} \tag{4}$$

The exponential distribution has special properties. If the component has survived to a time t_o, we will now verify what the probability is that it will survive a further time t'. Let this probability be given by $R(t_o + t' | t_o)$.

Then $R(t_o + t' | t_o) R(t_o) = R(t_o + t')$

$$R(t_o + t' | t_o) = \frac{R(t_o + t')}{R(t_o)} = \frac{e^{-\lambda(t_o + t')}}{e^{-\lambda t_o}} = e^{-\lambda t'} \tag{5}$$

The probability of surviving a further time t' is still given by an exponential distribution. For this distribution it means that the component has 'no memory'; a component that has survived to a time t_o, has the same chance of survival as a brand new component. The exponential distribution is the ONLY distribution that has this property. It has important significance with regard to replacement and maintenance policies. If a component has a constant failure rate, a brand new component is no better than an unfailed component in use, however long that component has been in service. In that case, preventive maintenance cannot be justified.

5 THE MTTF (MEAN TIME TO FAILURE)

The mean time to failure can be defined as the average time interval to the failure or in case of repairable systems the average time interval between two successive failures (MTBF).

The MTTF can be obtained by calculating the first moment of the failure density distribution or the mean value. After some calculations, following expression is obtained:

$$MTTF = \frac{1}{\lambda} \tag{6}$$

MTTF is often used in reliability studies, but one has to be careful when using this value, as will be shown in next example.

Example: Consider a device with $\lambda = 10^{-3}$ failures/hour or an MTBF, which equals 1000 h.

Suppose that we have 100 units, how many will be working than after 1000 h.

Using Equation (4), and setting $\lambda = 10^{-3}$ failures/hour and $t = 1000$ h, one finds:

$$R(t) = \frac{1}{e} = 0.367$$

or

$$N_w = N_o * 0.367 \cong 36$$

After 1000 h, only 36 units are still working. Figure 3 shows us the graphical interpretation, while Table 1 gives us the evolution of the reliability as a function of the time.

If the MTTF equals 1 year, the reliability values of Table 1 are obtained. $F(t)$ is the unreliability, and equals $1 - R(t)$.

As mechanical engineers, we are not pleased with the exponential model, where the commonly used preventive maintenance strategy cannot be justi-

Table 1. Reliability of a device with MTTF = 1000 hours.

Time	Reliability	Number of working units
1h	0.999	99
1day	0.976	97
1week	0.71	71
1month	0.486	48
1000 h	0.367	6
3 months	0.069	6
1 year (= 8760 h)	$1.5\ 10^{-4}$	0

Table 2. Reliability of device with MTTF = 1 year.

t	$R(t)$	$F(t)$
0	1	0
1 month	0.9200	.0800
3 months	0.7788	.2212
6 months	0.6065	.3935
1 year	0.3679	.6321
2 years	0.1353	.8647
3 years	0.0498	.9502

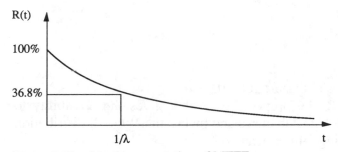

Figure 3. Graphical interpretation of MTTF.

fied. Moreover our engineering experience learn us that wear and fatigue are part of our daily life.

To take these phenomena into account, an increasing failure rate should be taken into account. A Weibull distribution is capable to present whatever evolution of the failure rate.

6 A MORE REALISTIC MODEL FOR MECHANICAL SYSTEMS: THE WEIBULL DISTRIBUTION.

In this Weibull distribution, the failure rate λ can be increasing, constant or decreasing depending on the shape parameter n. Every stage of the bath-tub curve can be modelled.

The failure rate takes following form:

$$\lambda(t) = \frac{nt^{n-1}}{\tau^n} \tag{7}$$

n is the shape factor, τ the time parameter.

Going back to the definition of the failure rate gives, after integration of Equation (3):

$$N = N_o e^{-\int_o^t \lambda(t)dt} \tag{8}$$

and

$$R(t) = e^{(\int_o^t \lambda(t)dt)} = e^{-H(t)} \tag{9}$$

If one substitutes Equation (7) one gets for $H(t)$

$$H(t) = \int_o^t \frac{nt^{n-1}}{\tau^n} dt = (t/\tau)^n \tag{10}$$

and for $R(t)$

$$R(t) = e^{-(\frac{t}{\tau})^n} \tag{11}$$

We supposed that failures only start at $t = 0$.

As shown in Figure 2, the proper choice of n gives the possibility to model every stage of the bath-tub curve and makes the Weibull distribution a flexible tool in modelling failure behaviour.

For $n = 1$ failure rate is constant (exponential distribution); $n > 1$ failure rate is increasing (wear-out); $n < 1$ failure rate is decreasing (reliability growth or work hardening).

In the case of the Weibull distribution the MTTF equals:

$$MTTF = \tau\, \Gamma(1 + \frac{1}{n})$$

where Γ is a gamma function.

Table 3 gives the value of MTTF as function of n.

If again a similar example as used, for an exponential distribution, is worked out for $n = 3$ and the same MTTF, namely 1 year, τ equals 1.1198 years and Table 4 is obtained.

Table 3. MTTF/τ as a function of *n*.

n	MTTF/τ
1	1.0000
2	.8862
3	.8930
4	.9064
5	.9182
6	.9275
10	0.9514

Table 4. *R(t)* and *F(t)* as a function of time.

t	*R(t)*	*F(t)*
0	1.0000	.0000
1 month	.9996	.0004
3 months	.9889	.0111
6 months	.9148	.0852
1 year	.4906	.5094
2 years	.0034	.9966
3 years	.0000	1.0000

The influence of wear-out can be noticed when analysing Table 4, and comparing Table 4 with Table 2. Up to MTTF, the reliability of the component is larger in the case of Weibull distribution (49.06% versus 36.79%). After $t = \tau$, due to the increasing failure rate, reliability will rapidly decrease (compare after 2 years).

7 RELIABILITY TESTING

For new materials or products, where no reliable failure data are available, we can only rely on reliability testing. During testing, all relevant parameters which could have a possible influence on the component's life, will be taken into account: temperature, humidity, vibration, corrosion e.o. Even for mechanical components, accelerated testing is required for obtaining reliability data in a relatively short time. Whereas for electronic reliability, failure accelerating rules such as Arrhenius law, can be of invaluable help, the correlation for accelerated reliability tests on mechanical components is not obvious at all. If moreover we are dealing with advanced and new materials such as composites, it demands extended research and testing, as will be shown in the next chapters.

8 CONCLUSION

Reliability is gradually becoming an important parameter in designing, producing, selling and bying products. For new and advanced materials, first of all reliability has to be proven. This is the main purpose of this book on durability of composite materials.

REFERENCES

[1] O'Connor, Patrick D.T. 1991. *Practical Reliability Engineering*, 3rd edition, Wiley, Chichester.
[2] Kapur, Kailosh C. & Lamberson, L.R. 1977. *Reliability in Engineering Design*, Wiley, New York.

Structural failures – Reliability: Learning from failures, and legal consequences

H.P. Rossmanith
Institute of Mechanics, Technical University Vienna, Austria

ABSTRACT: This contribution focuses on some of the main aspects of the interaction between engineers, scientists, lawyers and technical insurers within the framework of failure analysis, failure assessment and failure prevention in various fields of engineering. It is found, that learning from failures is a common though very expensive method of improving structural design and fabrication. In addition, in spite of the increasing interest of engineers in legal matters and the developing interest of legal experts in technical issues, there is still a wide gap between these groups of experts and the search for a common language and possible ways of effective cooperation is underway.

1 INTRODUCTION

It is generally accepted that various defects in materials and structures as well as shortcomings in the manufacturing processes may lead to premature failure of structural components and plants. These failures very often are associated with heavy loss of material and occasionally also lead to the loss of lives. The spectrum of accidents in all fields of engineering from locomotive axles to microelectronic switches and biomechanical applications is almost unlimited. Recently, failures in composite structures have received increasing interest as expressed in the appearance of a rather large number of papers and books in the literature [1].

Insufficient window reinforcement was the prime reason for the failure of the deHavilland comet airliner in the 50's of this century when several comet airplanes disintegrated en route in mid-air. At the close of the 20th century we take it for granted that science and technology advance and improve with time and contribute greatly to human achievements in creating a satisfactory environment for humans and life in general. Creation of technical artefacts and realisation of a technical idea also incorporates the latent

11

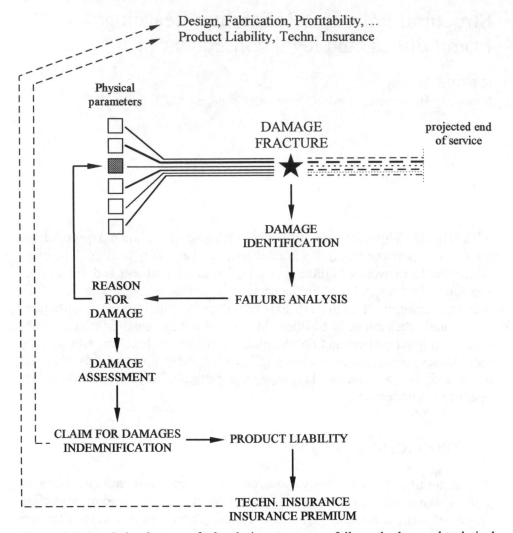

Figure 1. Interrelation between faulty design, premature failure, the law and technical insurance.

danger of not having paid attention to a crucial fact which later on may become the source of the evil.

If, in the logical chain from the design of a structural component to its production, use, maintenance and service as well as its recycling there occurs a 'mistake', this faulty procedure may terminate in a severe reduction of the component's or the entire plant's service life and lead to premature failure of the technical system. This is shown in Figure 1 [2] where one of the crucial design aspects has not been taken into account and a reduced service life can be noted. Failure of a technical structure or system very of-

Table 1. Industries reporting over 30 fatal injuries to employees: 1986/1987-1990/1991 (taken from [3]).

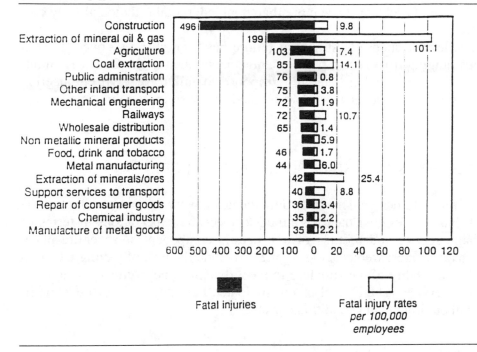

	Fatal injuries	Fatal injury rates per 100,000 employees
Construction	496	9.8
Extraction of mineral oil & gas	199	
Agriculture	103	7.4 (101.1)
Coal extraction	85	14.1
Public administration	76	0.8
Other inland transport	75	3.8
Mechanical engineering	72	1.9
Railways	72	10.7
Wholesale distribution	65	1.4
Non metallic mineral products		5.9
Food, drink and tobacco	46	1.7
Metal manufacturing	44	6.0
Extraction of minerals/ores	42	25.4
Support services to transport	40	8.8
Repair of consumer goods	36	3.4
Chemical industry	35	2.2
Manufacture of metal goods	35	2.2

ten is associated with heavy financial loss due to reduced production capability or even complete lack of production.

Damage assessment and indemnification will follow suit with legal and insurance-related aspects to be considered next. This interrelation between technology, law and insurance should of course then be characterized by a strong feedback to the designer informing the technical expert about 'What went wrong?'.

Very often the lesson to be learned from the investigations of failures is that a careful design in fact does reduce the risk of structural failure as a result of a material failure. This is the more important when the health and safety of humans are at stake in a wide range of industries, at home or during pastime. Death rates per hundred thousands of workers and injury rates per thousand employees have shown a monotonic increase during the past 50 years. Table 1 taken from the Plan of Work for 1992/1993 of the UK Health and Safety Commission (see [3]) gives some indication of the scale of the problem by showing a statistics of fatal injuries to employees in various industries with the structures and components involved in the accidents varying widely in size, material and method of fabrication.

Likewise, the total cost of accidents has increased exponentially during the past 30 years and concomitantly, insurance costs have risen accordingly. Over the period of 10 years (1971-1981) the total economic losses from work-related accidents have more than doubled, social inflation already corrected for.

The legal response to structural failure has been changing over the centuries ranging from severe punishment as mentioned in the Codex Hammurabi [4] the oldest legal liability response to failure of civil engineering constructions, to rather weak and relaxed response in certain societies and ages.

2 DESIGN AS A POSSIBLE SOURCE OF FAILURE

Failure due to design defects or deficiencies is very common and it can be said that the designer then has failed to meet the expectations in terms of reliable service and lifetime of a structural component or an entire plant. The aspects most frequently 'forgotten' include the lack of paying attention to the possibility of the structure to meet dynamic and cyclic stress loading during service, a high level of unanticipated humidity, acid or other detrimental environmental conditions or else.

2.1 *What is a defect?*

Even in our technologically advanced society there is still poor understanding of the issue of 'What is a defect?'. In fact, the defectologic terminology in structural failure analysis [5] to be understood by the engineers and clients, the public and media, insurance companies and the legal experts likewise still lacks adequate development.

Experience tells that there is hardly a constructed facility which does not either contain a flaw in its design or develop a flaw during its service lifetime. Economic pressures normally accompany if not dictate and control the road from design, manufacture, service, maintenance, recycling and deposit of a product, be it a consumer item, a structure or an entire plant or system. The term product here ranges from metropolitan transportation systems to systems of codes and requirements. The door for the initiation of a factual flaw which may lead to final failure of the product is open at each of these stages.

A distinction has to be made between defect, damage, fault and failure. A defect is considered the cause of the deviation of a structure from the path of expected and requested performance. Damage may be considered a partial loss of function of a structure which can be repaired to regain its fitness for purpose and active service. The definitions of a fault and a failure are of a somewhat fuzzy nature. Faults can be classified either as reversible faults, where the faultless or pre-failure state can be regained and the fault will disappear without substantial measures to be taken, or irreversible faults which are linked to the ultimate failure of the product. It seems clear that a high subjectivity of fault observation and assessment prevails.

2.2 *Design defect*

In the legal field there is no general definition of a design defect per se, however, a statutory definition for product defect may be found in the Second Restatement of Torts, 1965 [6, 7]:

'...who sells any product in a defective condition which is unreasonably dangerous is subject to liability for physical harm caused to the ultimate user if the seller is engaged in the business and the product reaches the user without substantial change in manufactured condition.'

In Austria, the Federal Statute of January 21, 1988 on the Liability for a Defective Product defines a defect as follows [8]:

'A product is defective when it does not provide the safety which a person is entitled to expect, taking all circumstances into account, with special regard to (1) the presentation of the product, (2) the use to which it could reasonably be expected that the product would be put, and (3) the time when the product was put into circulation. A product shall not be considered defective for the sole reason that a better product is subsequently put into circulation.'

Technically, a structure or a structural component suffers from a design deficiency or defect when the artefact cannot meet the expectations set by the customer and user at the design phase. Most of the design deficiencies can be related to lack of consideration of proper loading of the structure or structural component, improper selection of the material and heat treatment as well as stress relief, etc.

2.3 *Safety and risk of a design*

The safety of a designed product or system depends on various factors including and concerning conditions and activities associated with the use of the product or the system. The overall safety and risk of a product depends on the level of risk admissible and on the valuation of safety in a given situation. A decision will be made jointly by the designer, manufacturer, customer, user, regulation agency and quality control where trade-off situations arise with respect to cost, time, available material and technology and other important constraints and requirements – sometimes more important than safety [9].

Failure analyses have shown that safety is an all-embracing and general issue and cannot be left solely the designer's task. Input from many sides are involved: designer, manufacturer, user etc. However, with the greater responsibility of the designer with respect to the tailoring of the products to the individual needs of the customer, the designer also from a legal liability point seems to be made more and more responsible with respect to occupational safety.

With the Machine Safety Directive EN 292 [10] a means to incorporate safety considerations into the design process has been created though it must be complemented with comprehensive risk management features. It addresses risk reduction by design, safeguarding against unavoidable hazards including additional precautions against emergency situations and information and warnings to ensure the safe use of the product.

Proper design, material selection, the use of warnings and safety instructions will contribute to the safety of workers and users of the product, signalize the product fit for purpose and reduce the risk of accidents and limit legal liability [11].

3 MANUFACTURING AS A SOURCE FOR FAILURE

Deficiencies may develop due to faulty manufacture and imperfect assemblage as well as due to inadequate service and maintenance even when the product has been correctly designed. In products manufactured to specifications the defects can be introduced when certain requirements are not met. Careless changes in material specifications and unsatisfactory welding and joining of parts of engineering structures are among the most frequent post-design modifications – often carried out by unqualified or inexperienced personnel – which can be the cause of unexpected failures.

4 MAINTENANCE, WORKING PRACTICE, MATERIALS HANDLING AND INSPECTION

Many operational failures are attributed to defective material conditions, but upon closer examination the real cause of the defect can often be credited to maintenance related events.

Most engineering components, structures and entire plants require regular maintenance and inspection to guarantee proper functioning during their projected lifetime. Failure to maintain quality and negligence on the part of inspecting critical parts of the structure in many cases may lead to structural failure of a part of or the entire engineering structure. The importance of written instructions and the existence and workability of quality control during all stages from the design to inspection during service cannot be overestimated.

Maintenance related product failures can be classified by their basic causes [12]:

– Maintenance worker related failure;

– Design defect which makes maintenance difficult, unsafe or even impossible;

– Failure by the worker to recognize the consequences of 'short-cut' maintenance, 'routine' maintenance or modified performance of maintenance due to ignorance or limited understanding of fundamentals facts; and

– A combination of the above causes.

With the increasing efforts to economize on energy, material, design and labour expenses, which may exert a detrimental effect on the lifetime and performance of products, increasingly more problems will arise in connection with proper maintenance, repair and rejuvenation practices. Particularly in those countries, where for economic reasons plants and machines will be used for time intervals far exceeding their guaranteed lifetimes and under conditions of almost complete neglect of maintenance and safety precautions, questions about residual lifetime and safety attain high priority.

In many practical examples the execution of proper maintenance procedures (such as periodic cleaning and painting to minimize the deterioration of a structural product) will render a structure safe and guarantee the expected lifetime of the product.

Improper use of working tools and the adoption of unsafe working practices can be identified as the primary causes of industrial accidents [13]. In the field of material handling (loading, lifting, transporting, storage, unloading) the conditions influencing safety and risk are: equipment, material to be handled, environment, personnel and working practices. Most accidents are a chain of events with the final failures as the outcome but with one particular element to be identified as the principal source to initiate the accident.

The employment of unskilled labour and workers instead of craftsmen or experts frequently is a source of fatal accidents. Improved (on-the-job) training and education of staff, employing qualified personnel knowledgeable of applying the appropriate tools, and the enforcement of concurrent quality control should help reduce the risk of structural failure due to manufacturing.

Failures are frequently induced by faulty (incorrect sequence of operations and tasks to be executed) or incomplete procedures, manuals and installation instructions. Also, human error constitutes as a main source for accidents by allowing for excessive risk, improvisation on the basis of unfit working tools and working methods, etc.

Warnings serve to indicate potential problems. Manufacturers do have a duty to warn of potential hazards that can be anticipated by the ordinary user. As there is no simple definition of an 'ordinary user', foreseeing potential problems and dangers on the side of the producer is a 'must'. However, a warning is effective only when the message is received and understood and the user makes it the basis of his/her actions [14].

5 HUMAN FACTORS ENGINEERING AND LOSS CONTROL

Over the past decades human factors engineering or ergonomics has developed from a truly esoteric idea to an important field of engineering. Unfortunately, as with all new disciplines, it is not at all fully understood and appreciated by the designer and engineer, and often even less accepted by the consumer. When the interaction between people and the product gives rise to friction then the industrial designer has failed; in the contrary, if human lives are made safer, more efficient and more comfortable or happier, the designer has done a nice job. As legal liability is affected by technical and human errors, good quality design and excellent workmanship in manufacturing a product help in reducing potential losses and consequently in cutting down insurance claims [15].

Human error has been identified as a prime factor to product failure or loss of serviceability.

The assessment of the safety of a technical product rests on both, the analytical assessment of a safety factor as required by the design codes, rules and standards to account for uncertainties in the loading, material, manufacture etc. environment, and also on the effectiveness of the execution of the design, documentation, construction and use. It is clear that ignorance, negligence, carelessness and lack of knowledge on the part of those humans associated with the design, manufacture, use, service, maintenance and repair form the major causes of human errors. The relatively poor understanding of error occurrence and its significance for structural failure and its social and economic consequences has led to an upswing in ergonomics and human factors engineering [16].

From observations human errors in the design process and in the execution of the design are distinct events which include the incorrect choice of material, the omission of a detail etc. and range from errors of omission (failure to perform a certain task), errors of commission (incorrect performance of a task), sequential errors and extraneous acts to time limit errors (failure to perform a task within a certain time limit) and others.

Structural abuse by sabotage or acts of war are generally considered unpredictable and there is not normally an expectation of structural performance under these severe and unpredictable conditions. An exception though may be given by the design of nuclear power plants and high arch dams where the possibility and the effects of terrorist attacks and the likelihood of a plane crashing into the structure have to be taken into account in the design process via structural failure probability.

6 FAILURE ANALYSES: SOCIETAL AND EDUCATIONAL IMPLICATIONS

Primarily, failure investigations and analyses are conducted by experts to find the cause of failure of a product and – even more important – to derive measures for the prevention of these failures.

This immediately raises the question: 'Who is an expert?'. A possible definition characterizes experts as individuals with experience in the relevant field associated with the problem investigated. A technical expert, therefore, is known by his/her involvement in and commitment to a particular field of technology. Technical experts may be selected on the basis of their level of experience and competence in evaluating previous failure cases and their experience in the related industry.

Expert opinions are very likely subjected to biases which can be either motivational or cognitive and appear in the form of discrepancies in the results of design assessments and failure investigations. Motivational biases rest on the expert's desire to change the outcome of an investigation to his favour, whereas cognitive biases depend on the expert's mode of judgement often influenced by most recent occurrences of the event of interest [17].

Hardly ever recognized, experiences derived from failures form the main base for design improvements. The failures actually do not have to occur in service, in 90% of all cases product defect evolution can be observed in a preliminary – and often remediable – stage either during laboratory testing or in the field. Not only are designers forced by failures to redesign the products by changing certain characteristics of the product, but failures contribute importantly to the development of advanced materials testing and in modifying or developing new materials so that similar future breakdowns can be avoided [11].

Legally, the designer ultimately is made responsible for design defects. No matter how thoughtfully and carefully he/she designed the product, even with the best trained and competent design expert a mistake or oversight may go unnoticed. Formal design review procedures in which special emphasis is given on all foreseeable possible uses or applications of the product will help to alleviate the job and the stand of the designer.

Publishing information about failure analyses is a significant contribution to achieve these goals. In the relevant journals it is important to publicise and discuss the techniques applied and the outcomes of failure investigations including a number of comments as to 'what went wrong?' and 'why did it go wrong?'. The experience gained during failure examinations should be available to all experts concerned with this problem. This can be achieved by making this information optimally applicable by means of advanced computerized data storage.

Major incidents and accidents where public health, environmental or other issues are at stake and which very often are of public interest – also in the taxpayer's interest in spending her/his money wisely – should be discussed openly and the results of the investigations should be made public, in addition, information derived from failure investigations should be disseminated in publications such as information brochures, press releases and contributions in popular science and technical journals and weeklies.

This information, however, also must be studied and utilized to complement and improve academic qualifications in the form of continuing education and training. It is mandatory that the lessons learned from industrial failures be incorporated in educational programs and effectively put to work via feedbacking this body of valuable information. Reducing the risk or perhaps complete avoidance of failures can be attributed to a great deal to improvements in learning and training of personnel and staff to acquire a higher degree of competence and qualification.

Getting the message across from experts to workers is a very difficult task and education of the workforce may be best done by the implementation of company-sponsored programs. Acceptance of these educational efforts by the workers is strongly enhanced if they get the feeling that the incentive for further education, often 'on the job', originates in their ideas. For the benefit and safety of all, workers and consumers as well, there must be more education, greater knowledge and expertise and understanding of the operations, be it design, manufacture, assemblage, maintenance, service, repair and possible recycling. There is no better strategy for putting out useful, reliable and safe products than the feeling of and having pride in doing a good job.

7 FORENSIC ENGINEERING

The following considerations are primarily valid in the US law scene and in US court rooms.

Forensic Engineering has evolved as the art and science of investigative engineering, i.e. determining the facts of a case. It is a field of engineering where technology and the law meet. The major areas of forensic work are: product liability, accident investigation and reconstruction, and contract liability [18].

The tasks to be performed by a forensic expert include to determine:
– 'What happened?

- 'Where did it happen?'
- 'How did it happen?'
- 'Why did it happen?'
- 'Who caused it to happen?'
- 'Who is responsible for what happened?',
- 'What are the cost to repair or replace?' and
- 'What are the damages suffered by various parties involved?'.

7.1 *The forensic expert*

A 'forensic expert' may be defined as a highly trained specialist – often an engineer or scientist – in the field addressed, who has no specific ties to and interests related to any of the litigants in the subject case other than being engaged to investigate and to render an objective professional opinion (including the approach and conclusions) regarding the causal aspects of the case [18].

A forensic expert may be defined as a person entrusted with solving all the problems listed above single-handedly.

The professional profile of a good forensic engineer must yield a technical expert equipped with excellent investigator capabilities, a superb detective with a high level of engineering intuition for being able to keep the attention and excitement of the judge and the jury at top level, and a man of courage ready to enter dangerous ground while still standing the man/woman with calm and cool temper when confronted in the attorneys' cross examination.

Modern litigation makes full use of the expertise of various experts; injury cases routinely use economists, personal injury cases use medical doctors and therapists, product liability and transportation accidents use reconstruction experts and construction cases regularly require engineers and architects.

In the USA the legal definition (Federal Rules of Evidence 702) of an expert is any person who possesses specialized knowledge through 'knowledge, skill, experience, training and education' [19]. This definition is too broad and the Federal Rules of Evidence allow an individual much latitude in offering forensic engineering services and virtually everybody to testify as an expert.

In contrast, an 'in-house expert' possibly with the same background and qualifications and employed by one of the litigants is generally considered to have a bias toward his employer's viewpoint.

8 TOTAL QUALITY ENGINEERING AND RISK ENGINEERING

Regarding industrial products the main concern of engineers must be safety and quality including control of by-products of production, manufacturing processes, service and possibly recycling or waste disposal.

8.1 *Safety and risk*

With the requirement of more safety and the imposition of environmental legislation and regulations as well as the impact of the insurance industry, engineers, therefore, have changed their attitude and have been forced to learn to act in a manner responsive to both the society's constant demands for new and advanced products and requests for increased safety. Nevertheless, the international market is full of junk products that harm the consumer and user (e.g. toys with highly toxic paint that injure children, etc.) and the designer and manufacturer is constantly being pressed by the consumers' desires for inexpensive products ranging from paper clips to aeroplanes. Interestingly, when the consumer's and designer's welfare is at stake, then safety and quality rank first and the lure of the cheap remains ineffective.

The main goal of risk engineering is risk reduction, i.e. taking measures to avoid or at least reduce losses incurred to anyone in the chain from designer to consumer. Risk engineering comprises identifying and controlling hazards, eliminating unacceptable risks, the implementation of risk reduction measures, etc. by applying engineering knowledge. As many insurance companies now perform and offer risk engineering activities and services [20], respectively, and all risk engineering efforts are directed toward safer and better products, excellent co-operation between the risk engineer – often an insurance expert – and the technical engineer is mandatory. Insurance companies in all major industrialized countries have already established or are in the process of establishing national and international global networks of risk engineering centers where professionals with excellent technical, insurance and legal expertise and reputation offer and provide a wide range of interprofessional and interdisciplinary technical, insurance and legal synergic services.

When risks eventuate, questions on responsibility and liability will arise and assessment for damage and loss becomes inevitable. The questions 'which party can best foresee, control, and bear the risk, and benefit or suffer due to eventuation of the risk' form a basis of fairness in the allocation of risk to the various parties involved in a failure case, and they need detailed answers. Risk and liability issues with respect to contractual and tortuous liability as well as the field of insurance for construction risks is highlighted in Ref.[21].

8.2 *Repair and liability*

Particularly susceptible to the issue of risk engineering is the current trend among companies all over the world to increasingly postpone the building of new installations and facilities such as e.g. power stations, port facilities,

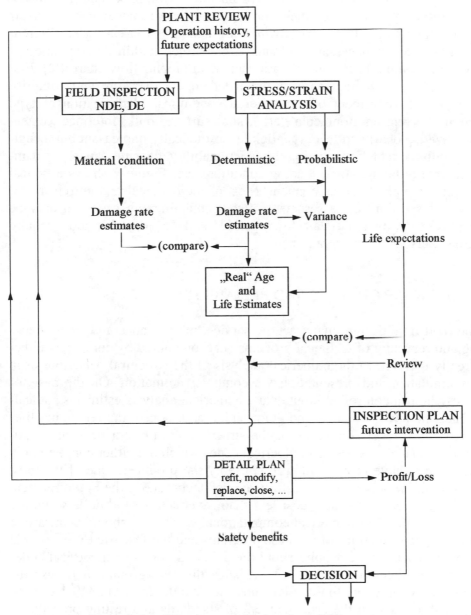

Figure 2. Flowchart of plant life assessment (taken from [22]).

petrochemical plants etc. in favour of rejuvenating and modernizing exis-
tent older plants and facilities. Escalating cost, lack of foreign currency, ban
on supplies, low credibility on the international monetary funding sources,
reduced volume of production, delay of capital investment and other reasons may be made responsible for this attitude and reluctance to modernize.

Life extension of existing frequently old machinery, plants and facilities
has become a big and often lucrative business. Complex facilities do not
age uniformly throughout their service time, and replacement or repair/
refurbishment of selected parts which are ageing and wearing at different
rapidity becomes necessary and can be performed within a systematic approach according to a plant life assessment and repair flow chart [22]. Elements of a systematic approach are listed in the literature and comprise
methods of assessment of the material conditions, identification of type
(fatigue, creep, environmental, etc.), site (surface, bulk, interface), extent
(localized, global) and level (negligible, aesthetically unpleasant but technically unimportant, endangering the functionability, causing danger) of damage, state of temperature, loading conditions etc. Figure 2 shows a typical
flowchart of plant life assessment. Most of the classical and current methods of inspection and assessment of the remaining life time can only be
employed during plant shut-down periods and the available tools require
skilled personnel for operation.

9 SOFTWARE QUALITY AND RELIABILITY

The road from design, manufacture, service, maintenance and even wreck-
ing and recycling of technical products is accompanied by computers either
directly involved in the numerical analysis of the structural behaviour or in
the simulation and presentation via computer animation. On the basis of
mostly human controlled input data a numerical analysis estimates and calculates the levels of stress and deformation and, hence, the remaining lifetime or load level admissible of the structural component or a plant. Numerical simulations base on numerical codes which are either commercially
available or have to be written in-house by the producer. One of the most
famous and widely and successfully applied technique is the Finite Element
Method. Current programs cost less than several thousand dollars and are
routinely utilized in numerical computational work. One should be aware of
the possible limitations of all these programs and the forensic expert should
be able to make a clear point about the do's and dont's of numerical codes
and the applicability thereof [23]. Against this background of unease the
National Agency for Finite Elements and Standards NAFEMS has been
founded to assist in the establishment of standards and testing procedures,
publishing of education and training requirements, finding a common plat-

form for users and developers, and in maintaining a register of FE systems and users [23]. With respect to product liability any software is a general product [24].

The availability of increasingly powerful computer hardware and the ever-increasing number of software companies on the market as well as the sheer incomprehensible number of software programs available even to the layman user, the immense pressure to reduce development time and other human factors, software quality [25], software reliability and defect prevention in the course of the development of software have become prime issues also in court cases.

Modern structures depend on numerical computations and the dominant role of finite element methods as a design tool and in major analyses of increasingly large and complex structures has drawn the attention of risk and reliability engineers and managers to software quality and safety. While the additional information gained from numerical simulations undoubtedly assists the designer and manufacturer and enriches the knowledge of a structure's behaviour under various service conditions, the crucial question posed by all assessors, clearance and certifying authorities is concerned with the degree to which an engineer/designer would trust such an analysis as the sole method for assessment of safety, particularly in cases where the safety of humans depends on the structural integrity.

In the use of numerical methods for structural clearance the certifying and clearance agencies must have confidence in the procedures used to control the analysis function, in the modelling practice used for analysis, in the integrity and reliability of computer codes, in the capabilities of the personnel responsible for analysis, and in the authorities' own ability to assess analysis-based procedures [23].

One of the serious difficulties is based on the fact that software quality is only indirectly measurable and the most commonly encountered defect causes refer to [26]:

– Education (e.g. the function to be implemented, the tool or program language used etc. has not been understood);

– Communication (lack of communication within the team);

– Oversight (not all possible cases were considered);

– Transcription (the operation to be performed is known but a mistake is made in the execution); and

– Process (error due to a flaw in the development process).

Defect prevention within the framework of software should start with a management commitment to the process, the demonstration of how to prevent a defect to the whole development team, and the nomination of an action team [27].

10 LIABILITY FOR DEFECTIVE PRODUCTS

Product liability and product liability legislation focus on consequential damage (property, health, life) incurred to and suffered by animate or inanimate objects in the course of another event. The question 'Who is liable for damage resulting from a defective product?' is paramount in product liability [28].

Product liability, however, does not deal with issues of warranty, removal of a deficiency of a product or compensation for a deficiency, nor does it address the damage inflicted to the product itself.

10.1 *Legal notion of a product*

Although the term 'product' is very difficult to define, it often includes all man-made movable and stationary items, single or possibly incorporated into a movable or immovable assembly product characterized by a definite time when the product was put into circulation and service. Under Austrian law, electrical energy is considered a movable product, whereas agricultural items become products once they have been processed.

A defective product which does not provide the safety the user or consumer is entitled to expect must be distinguished from a product which lacks an attribute generally associated with the product and therewith violates the law of warranty and fails to meet the expectations expressly promised in a contract. A defective product obviously lacks an attribute that's reducing and modifying safety.

From a legal point of view a product may become defective already at the stage of advertising for it. Make-up, insufficient or even faulty explanation supplied to the user can render a product 'legally defective'. In a product liability case the instant of time is important when a product has been released into circulation by buying, renting, leasing or otherwise and thereby the producer has lost the possibility to control a possible risk involved in the use of the item. Being outdated at the time of circulation by the sole cause of availability of a better suited product is not considered defective.

10.2 *Who is a producer?*

The legal term 'producer' extends far beyond its colloquial meaning. Generally, the notion 'producer' is associated with a person who identifies himself or herself as the producer by putting a distinguishing feature (design, fabricate, select material, manufacture, assemble, attach a trade mark or a strange name to the product thereby hiding the real origin of the product etc.) to the product.

Historically, modern product liability legislation bases on the idea that the producer of a product found defective can cover the economic risk easier than the third person (user, consumer etc.) suffering from the damage. This is in line with the departure from the tendency of covering one's own expenses due to damage toward (often very successfully) finding somebody responsible and with deeper pockets for recompensation. Product liability does not result from a contract but from a delict committed by way of circulating a defective product. It is not only immaterial whether or not the injured person or the owner of the property damaged are in a contractual relationship with the manufacturer or the importer of the defective product, but responsibility is established by the sole existence of the defect rendering the product unsafe independent from any guilt (e.g. Airlines-Act, Pipelines-Act, Nuclear Liability Act, Water Right Act etc.).

10.3 *Finding a scape goat*

Liability can be passed on from the supplier to the importer and back traced in a chain where several businesses are involved by identification of the relevant predecessors within a reasonable time. This regulation of 'strict liability', which is a 'no-defect liability' is mostly acknowledged and applied in the USA (Second Restatement of Torts Art. 204A). With the 'Council Directive of 25 July 1985 – 85/374/EWG – of the European Community the product liability scenarios in Europe and the USA have been partially harmonized as there are still substantial procedural differences between the two legal systems which might lead to widely differing outcomes when applied. The situation in Japan where liability with respect to a defective product rests on the Civil Code is very different but similar to European pre-EC regulation characterized by a lack of no-fault liability as well as compensation under punitive damages [28].

The existence of a multitude of possibilities to conduct a product liability case is of utmost importance when considering the complex entanglement of the international global marketplace dominated by technological products of highly supranational provenance. Damage claims associated with a defective product of multinational origin will quickly put the item of concern for discussion on an international level where the rules of private international law state that a conflict with extra contractual damage be governed by the legislation of the country in which the damage occurred.

Compensation for damage and loss (of property, health or life) is regulated by the Civil Codes where in the case of property damage the kind and extent of compensation depend on the trading value whereas in the case of bodily injuries to animate sufferers (humans and animals) appropriate compensation for physical and psychical pains can be asked for.

Joint or proportional liability and responsibility for a defective technical product may be shared by several persons in the genetic and generic chain from the designer, manufacturer, supplier, importer, distributor to the 'producer'. In the case of joint or shared guilt the damage or loss is caused by both, the defect in the product and the user e.g. by possibly inappropriate handling of a defective merchandise.

10.4 *Escaping strict liability*

Strict liability is meant to shut the escape door of exclusion of liability for a defective product. However, several cases of exclusion of liability are widely acknowledged [29].

These exclusions refer to cases where:

– A product, either stigmatized with a defect or developing a defect in the course of service, but complying with mandatory regulations issued by public authorities (law, decree, act, court decision etc.) would put the producer into a conscientious doubt or conflict between fulfilling a mandatory regulation and knowing about a deficiency of the product;

– The defect causing the damage or loss already existed at the time of circulation of the product but could not be discovered and identified as the source of malfunction on the basis of the state-of-the-art and standard of science and engineering; it is immaterial whether or not the producer had knowledge of the current state-of-the-art of science and engineering, the decisive fact lies in the possibility to discover and identify the defect by a scientist or engineer at the time of putting the product into circulation;

– The person under charge of negligence is the supplier of a structural element or part of a structural product, and the failure can be attributed either to the defective design in which the element or the component has been integrated, or to the faulty instructions supplied by the manufacturer of the product.

– The burden of proof is reversed;

– A non-private person who acquires the product for a professional activity will be excluded of liability by agreement.

10.5 *The role of the state-of-the-art in product liability cases*

The second of the above-mentioned cases of exclusions from product liability merits further investigation and discussion, as state-of-the-art relates primarily to design defects. The state-of-the-art defense in product liability cases seems to represent a highly controversial issue with numerous arguments in favour of and against it. The courts' inclination and reasoning ranges from the use of risk-utility analysis and customer expectation tests to determine whether the product was defective for strict liability purposes, to maintaining a policy of unbending definition of strict liability without re-gard to scientific or engineering knowledge [30]. Undoubtedly, the stiff position of courts is hoped to advocate and promote increasing research in safety research.

Producers must accept responsibility for complex advanced products which can be harmful without the users or consumers being able to ascer-tain the danger of the product or its use. A producer, therefore, who fails to keep himself/herself abreast with the modern development in science and technology and fails to warn of unreasonably dangerous conditions should bear the cost of injuries, damage and loss caused by the product. A con-sumer, generally, does not have the capability to recognize or even judge the level of danger when using the product. A later discovery of an insuffi-ciency should not be considered an indication of diligence. Producers must not be expected to be omniscient and include future aspects into the design and manufacture of products, and waiting for the ultimate in technology would not solve the problem as there will always be more advanced (and often better) products to come.

Many injuries and damages, however, are caused by actions which ren-der a product defective only when used inappropriately or in a different way as indicated in a warning.

With today's world of complex machinery and infinity of consumer products a public policy makes sense and is favoured, which puts the con-sumer into a position where he/she is powerless to protect himself/herself when using the product. There is, though, a basic problem associated with the practical execution as to the definition of a defect as a major obstacle and impediment to fair application. The question arises: 'When does the definition of a defect cease to hold and when does it become a definition of absolute liability'.

The state-of-the-art defense allows a manufacturer to prove that the de-sign was according to best knowledge at the time the product was launched on the market. In warning cases, where the inherent flaws or danger was discovered after the product had been released a seller/producer will be held liable if the danger could have been reasonably foreseeable and scientifi-cally discoverable.

11 CONCLUSIONS

In the complex world of the closure of the 20th century characterized by an unprecedented advance and progress in technology and a keen eye on opti-mality in business, technological advances – often a blessing for our society – are traded off against higher risks and, consequently, higher losses in the case of an accident.

As total risk avoidance may not seem feasible with many complex innovative industrial products, the engineer's and designer's duty lies in assessing and minimizing the risk within the frameworks of Engineering and Management of both, Total Quality and Risk. Proper design cannot be left solely with the designer and engineering considerations but direct or indirect input from other professional disciplines ranging from ecologists, sociologists, psychologists, risk managers, insurance analysts, politicians, educators and attorneys must be utilized. In fact, in national and international projects the whole community or governments of states ought to be involved in the planning of the project.

Interdisciplinary interaction of technical engineering, human factors engineering and product liability together with risk management and the insurance industry is required to guarantee the evolution of reliability based designs and manufacture of structures or facilities which meets current laws, incorporates concepts of safety, total quality and technical insurability.

Continued efforts of a large number of experts from the fields of engineering, law and technical insurance in narrowing the bridge between these professional groups and building a common platform for inter-disciplinary and interprofessional co-operation and understanding have led to the foundation of the International Society for Technology, Law and Insurance (ISTLI) with its headquarters in Vienna. This society aims at the promotion of international co-operation of technical, legal and insurance-oriented experts within the framework of structural failures and the technical, legal and insurance aspects thereof.

REFERENCES

[1] R. Talreja 1994. *Damage Mechanics of Composite Material*. Composite Material Series, 9. Elsevier.

[2] H.P. Rossmanith 1994. FPE – Failure Preventive Engineering: Fracture Mechanics Betwixt Designer and Failure Analyst. In: *Proc SPT-1*, pp. 11-21, Vienna, Elsevier.

[3] C.E. Nicholson et al. 1993. Common Lessons to be learned from the Investigations of Failures in a Broad Range of Industries. In: *Proc. SPT-4*, pp. 268-275, Vienna, Elsevier.

[4] B. Ross 1984. What is a design defect? In: *Proc SPT-1*, pp. 23-71, Vienna, Elsevier.

[5] D.W. Noel & J.J. Phillips 1992. *Products Liability in a Nut Shell*. West Publishing Company, 3rd edition.

[6] K. Haekkinen 1993. Are the safety measures taken by the designer adequate? In: *Proc. SPT-4*, pp. 1-8, Vienna, Elsevier.

[7] European Committee for Standardization, Safety of Machinery, Basic Concepts, General Principles for Design 1990. pren 292 Final Draft, Brussels.

[8]   U. Gramberg 1985. Influence of failure analyses in materials technology and design. In: *Proc. ASM Conference,* pp. 109-117, Salt Lake City.

[9]   C.O. Smith 1985. Maintenance related failures. In: *Proc ASM Conference*, pp. 333-339, Salt Lake City.

[10] M. Tichy 1991. From flaws to faults, *Structural Safety 9*, pp. 243-246.

[11] R.E. Melchers 1989. Human error in structural design tasks. *J. Struct. Engg.*, pp. 1795-1807.

[12] The Code of Hammurabi 1904. University of Chicago Press.

[13] G.A. Peters 1986. A safe approach to quality products. *Quality Progress,* January.

[14] C.O. Smith 1985. How much safety? Who decides? In: *Proc ASM Conference*, pp. 325-331, Salt Lake City.

[15] J.W. Juechter 1985. The expert witness and the attorney: vs or vis-a-vis. In: *Proc ASM Conference*, pp. 99-107, Salt Lake City.

[16] J. Mohammadi et al 1991. Evaluation of systems reliability using expert opinions. *Structural Safety 9*, pp. 227-241.

[17] English version of the Austrian Product Liability Statute 1989. *Economy,* 1: 61-62, March.

[18] J.S. Kirkwood 1993. Product liability – The search for new and deeper pockets. In: *Proc. SPT-4,* pp. 145-152, Vienna, Elsevier.

[19] B. Klikar 1993. Risk engineering. In: *Proc. SPT-4*, pp. 170-178, Vienna, Elsevier.

[20] B.A. Suprenant 1990. Forensic engineering – The good, the bad, and the ugly. In: *Proc. SPT-3*, pp. 239-244, Vienna, Forensic Engg 2, ½, Pergamon Press-Elsevier.

[21] R.K. Penny 1993. Some aspects of life assessment, inspection and monitoring of plants. In: *Proc. SPT-4*, pp. 487-494, Vienna, Elsevier.

[22] E.W. Green & P.F.Packman 1993. The technical-legal interface in presenting technical information in trials of fact in products liability cases. In: *Proc. SPT-4,* pp. 130-136, Vienna, Elsevier.

[23] W.M. Mair & A.J.Morris 1984. *The National Agency for Finite Element Methods and Standards.* NAFEMS WM99, September.

[24] W. Dette 1993. Software quality. In: *Proc. SPT-4,* pp. 227-236, Vienna, Elsevier.

[25] T. Kellermann 1993. Defect Prevention in Software Development. In: *Proc. SPT-4,* pp. 245-248, Vienna, Elsevier.

[26] M. Bartsch 1991. Product Liability for Software (Produkthaftung fuer Software). *Office Management 7-8*, pp. 15-17.

[27] T. Yamada 1993. On international patent law. In: *Proc. SPT-4*, pp. 34-38, Vienna, Elsevier.

[28] H. Bisanz 1993. Comparison of product liability law cases in EC, US and Japan – Status quo and beyond. In: *Proc. SPT-4*, pp. 28-33, Vienna, Elsevier.

[29] B. Moser 1990. Law concerning the liability for defective products. In: *Proc. SPT-3, pp. 307-308, Vienna, Forensic Engg 2, ½, Pergamon.

[30] F.P. Land 1993. The state of the art defense in product liability: a positive concept or an abomination. In: *Proc. SPT-4*, pp. 137-144, Vienna, Elsevier.

The abbreviation SPT-x stands for xth International Conference on 'Structural Failure, Product Liability and Technical Insurance'.

Application of composites in a marine environment: Status and problems

Peter Davies
Marine Materials Laboratory, IFREMER, Centre de Brest, Plouzané, France

ABSTRACT: This paper describes the current status of composite applications in a marine environment. After a brief introduction to this environment and the materials commonly employed, three types of application are described:
– Surface vessels;
– Offshore structures;
– Underwater applications.
Examples of each are given to illustrate the particular loading conditions which need to be considered. Constraints on the more widespread use of composites are then discussed, particularly with respect to certification of society requirements and fire regulations. This is illustrated by work performed on fire testing of panels and pipes. Finally the areas of active R & D at the present time and the needs of the marine industry (design guidelines, more efficient fabrication, long term property data etc.) are discussed.

1 INTRODUCTION

Fibre reinforced composite materials offer tremendous potential for applications in a marine environment, where their corrosion resistance and light weight are their principal advantages compared to metallic structures. Many applications exist and overviews are available [e.g. 1-3]. Considerable efforts have been made over the last 25 years to improve the understanding of the durability of these materials but design safety factors remain high for loadings other than static (long term, cyclic, impact). There is also a widespread mistrust of polymeric composites for fire-sensitive areas, in spite of considerable experience on passenger ferries in Scandinavia and increasing use offshore. This paper gives a brief summary of the current status of composites in marine applications and highlights some of the problems remaining.

2 MATERIALS

The materials which are being considered for the majority of marine applications are not the high performance carbon fibre composites, prepared by elevated temperature cure of prepreg layers, which have been adopted by the aerospace industry. Here we are mainly concerned with glass fibre reinforced composites prepared by contact moulding (hand lay-up). Typical fibre volume fractions are around 30-40%. There is only a little use of carbon fibres with epoxy resins and honeycomb core, confined to racing vessels and luxury boats where price is not an important parameter in design. For tubes and tanks filament winding or contact moulding are the main fabrication methods. Typical resins are polyesters, epoxies, vinyl esters and phenolics. The reinforcements are generally woven fabrics, often coupled with chopped strand mat layers. The ply-based analysis using laminate theory is therefore of limited use as unidirectional ply data are not available.

In addition to the monolithic composite structures there are also a large number of applications of sandwich structures. The most frequently used core materials are closed cell PVC foams and balsa. These typically have densities from 80 to 200 kg/m^3 but show poor fire resistance. Heavier mineral based cores may be the only solution when fire performance is critical.

3 THE MARINE ENVIRONMENT AND AGEING

The main threat to structures operating in a marine environment is usually perceived to be water, but more generally their durability may be reduced by:
  – Mechanical loads (wave impact, erosion, hydrostatic pressure);
  – Physical degradation (differential swelling due to moisture, thermal effects);
  – Chemical attack (hydrolysis of resin, effect of hydrocarbons);
  – Biological attack (fouling, biologically induced corrosion).
Environmental effects on composites have been widely studied [4,5]. Data have been collected for high performance aerospace composites, generally using varying relative humidities rather than immersion, while glass reinforced materials have been studied for chemical engineering applications. A large database has also been collected for naval applications, with over 20 years immersion in some cases [6]. One of the key issues in estimating long term ageing effects is the validity of accelerated test procedures. The use of increased temperature to accelerate testing times does not necessarily affect the different ageing mechanisms in the same way. The mechanisms which can intervene during ageing in water include the following:

Table 1. Aging mechanisms.

Reversible effects	Inversible effects
Plastification	Hydrolysis (molecular chain breakage)
Swelling	Leaching out of material
	Cracking and delamination

The time to the onset of hydrolysis is a critical parameter for durability predictions but few reliable data exist for commonly-used resins. The enhancement of degradation by applied stress has been examined by several authors and reviewed [7, 8].

A second type of degradation, which has been the cause of much controversy in the pleasure boat industry, is blistering. The appearance of blisters results from osmosis across the gel-coats used to protect composite hull structures. The phenomenon has been known for many years [9], and particular combinations of manufacturing conditions, resin chemistry, fibre coating and service conditions have resulted in blisters appearing in very short times. While initially an aesthetic problem, delamination and property loss may follow if blistering is not treated. Repair of blistered hulls can be very expensive as thorough drying is recommended. A recent study has examined the kinetics of blister propagation through accelerated tests. It was concluded that the probability of blistering appearing during the 20 year lifetime of a boat with orthophthalic polyester laminate and gelcoat was high, whereas for isophthalic polyesters this was much reduced unless the gelcoat was thin [10].

4 CURRENT MARINE APPLICATIONS OF COMPOSITES

4.1 *Surface vessels*

4.1.1 *Pleasure boat industry*
Small pleasure boats have been built from composites for over fifty years [11]. The principal fabrication route is hand lay-up, using glass/polyester composites, although there is some interest in injection methods such as RTM (Resin Transfer Moulding) for larger series. Competing materials are wood and aluminium but price and ease of maintenance have resulted in composites representing around 90% of the market. The technology involved in pleasure boat construction has been reviewed recently in a French national programme, the 'PNC' ('Projet National Composites', [12]). A significant innovation in this area is the growing awareness of the benefits of quality control procedures. In this context it is interesting to examine one

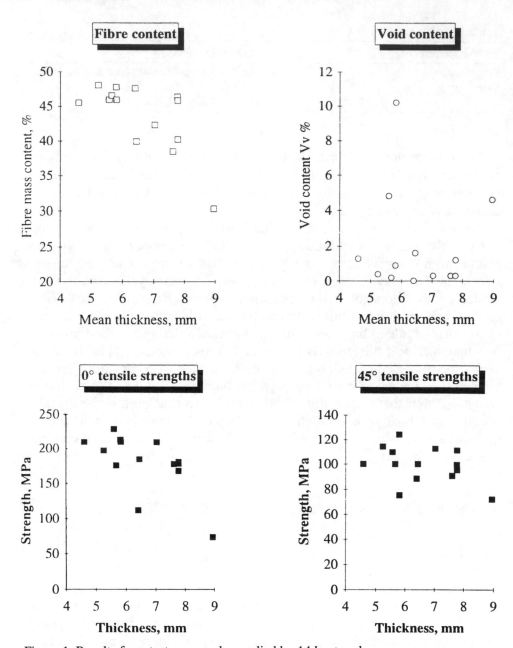

Figure 1. Results from tests on panels supplied by 14 boatyards.

of the myths concerning marine composites, which is that the hand lay-up process results in a huge scatter in material properties.

As part of the PNC programme 14 boatyards were asked to each supply a panel of glass rovimat reinforced polyester. Rovimat is a reinforcement

Table 2. Coefficients of variation of tensile properties for panels from 12 boatyards.

0° modulus	45° modulus	0° strength	45° strength
9%	13.5%	12%	10%

made up of layers of woven roving and chopped strand mat lightly stitched together. Each boatyard produced a one square meter panel consisting of 5 layers of rovimat. These were subsequently tested with both physico-chemical analyses (DSC, burn-off, void content), and mechanical tests (tensile at 0°, 45°, -45° and 90° and short beam shear) being performed. Over 700 specimens were tested. Some results are shown in Figure 1.

The interpretation of these results must be made with care, as some boatyards applied gelcoats of varying thicknesses and the fibres used were not always those requested so fibre contents and panel thicknesses vary. One material also contained additional fillers which strongly influenced 0° tensile strengths. Nevertheless, with one exception, due to a particularly thick gel-coat, the fibre contents varied from 38 to 48%. The void contents as determined by image analysis were less than 2% for all but three panels. There are always difficulties in obtaining reliable void content measurements but these values suggest that impregnation was reasonably consistent. Concerning the mechanical properties, stiffness is the main requirement of composites in most applications. If the material with 10% voids is excluded the tensile modulus in the fibre direction varied from 11.4 to 15.1 GPa while at 45° the variation was from 7.4 to 11.1 GPa. Strengths are also reasonably constant, with the exception of the very thick and the filled materials. The coefficients of variation about the mean for moduli and strengths are presented in Table 2 and appear quite reasonable.

4.1.2 *Passenger transport*

There are an increasing number of fast passenger vessels under construction and the design of such vessels will be used to illustrate the origins of safety factors in design.

Vessels transporting passengers in international waters are subject to SOLAS (Safety of Life at Sea) regulations issued by the IMO (International Maritime Organization) which severely restrict the materials options. For large ships the hull and most bulkheads must be non-inflammable, thus excluding polymeric composites. For smaller boats and fishing vessels the rules are less strict. The IMO tests are described in more detail in section 5 below. In Sweden and Norway sandwich construction is widely used for fast passenger transport and since 1986 sixteen surface effect ships (SES) passenger transport vessels have entered service in Norway [13].

Ifremer was involved in the design and construction of a surface effect

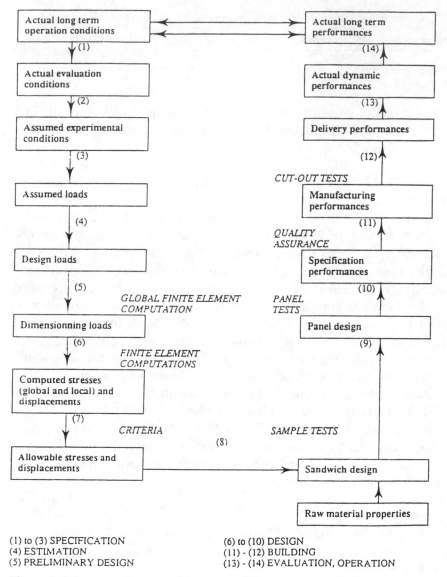

Figure 2. Schematic diagram of design/qualification process [15].

ship, the NES 24, built to transport 150 passengers [14]. The final design was largely based on quadriaxial stitched glass reinforced epoxy on a PVC core. Rolin has detailed the different partial safety factors involved in design, which highlight current areas of uncertainty in the use of sandwich composites for fast ships [15]. The design and qualification procedure is shown schematically in Figure 2.

Some comments may be added for the different steps:

1-5: Loading. The loads seen by the hull are quantified in terms of static pressures, whereas in practice slamming loading by waves can cause high core shear stresses or high local bending stresses in composites. This is still a grey area in design as insufficient data are available.

6: Calculations. Frequently finite element analyses are used to study global stresses and displacements. Test-finite element prediction correlationís on sandwich sub-structures have shown the difficulties inherent in such calculations, as will be described later (Section 6).

7: Admissible values/calculated maximum. For laminates failure criteria such as Tsai-Hill or Tsai-Wu are used, while for the areas subjected to slamming the maximum core shear stress is used. For the NES 24 a value greater than 1.5 was used for this factor.

8: Ultimate design values/admissible values. These factors are given by the classification society rules and for this case the factors were over 3.

9: Ultimate tested value/ultimate design value. A very large number of tests were performed. For sandwich materials considerable differences in properties can be obtained when different test standards are employed. For example the suppliers data for one PVC core's shear modulus varied from 16 to 29 MPa according to which test standard was followed.

10: Ultimate specified value/Ultimate tested value. A difference of 10% between initial estimations and measured values of weight of stiffened panels is not unusual.

11: Ultimate construction value/ultimate specified value. During the construction variations of up to 30% in the weight of the core supplied from the same source were noted. A rigorous quality control procedure is essential to limit this factor. Insert, restratification and other details must also be taken into account.

12: Final reception value/ultimate construction value. The variations in material properties overall measured on offcuts were less than 20% and varied on the safe side.

13-14: The influence of creep, fatigue, ageing and impacts must also be considered and few data are available. As more in-service experience is obtained these factors should decrease.

Overall these partial safety factors result in a total factor between initial estimate and final reception of greater than 5. There is clearly scope for reducing this coefficient but a major factor is lack of knowledge of the design loads for such vessels. The materials component can be reduced to reasonable proportions by strict quality control procedures.

4.1.3 *Fishing boats*
There are applications for composites in the structures of small and large fishing boats. Ifremer has recently been involved in an application of composite cooling water pipework in a fishing boat which illustrates very

clearly potential advantages and problems. In medium sized vessels most sea water cooling circuits are currently galvanized steel. The advantages of composite pipes are recognized, but the main obstacles to their more wide-spread use are higher cost and maritime regulations.

In 1990 the steel sea water circuit (which removed the heat from the engine cooling system) in a 59 m fishing vessel, was replaced with glass/epoxy pipework. The initial cost of the system was estimated to be 3 times that of the existing steel circuit, but the latter need replacing every 15 months with considerable labour costs. Four years after installation the composite circuit is still working satisfactorily [16].

4.1.4 *Military vessels*

The development of mine counter measure vessels (MCMV) has resulted in one of the largest single marine applications of composite materials. These vessels are typically 50 m long and the non-magnetic properties of composites have made them the first choice material in Europe. Smith has presented a concise overview of the different construction methods adopted since the first all GRP vessel, the HMS Wilton, was built in England in 1973 [1]. The way in which different countries have approached the design of these vessels is instructive and is summarized in Table 3.

In the UK transverse stiffeners on a monolithic hull have been used, as this was found to be an effective design against underwater explosion loading. The tripartite minesweeper (France/Belgium/Holland) was of similar single skin design although stiffener/hull joints were reinforced against explosion by GRP pins rather than by bolts. The Italian Navy favoured a thick hull to avoid problems with stiffeners, reaching 150 mm thickness at the keel. The Swedish navy, in collaboration with core material suppliers, developed the use of sandwich materials [17], and the Norwegian and Australian navies arrived later at a similar materials choice. Thus for vessels performing very similar roles completely different design philosophies have been adopted for different reasons.

Table 3. Designs of mine countermeasure vessels.

Country	Design	Class	Details
UK	Monolithic	Hunt, sandown	Stiffened
Italy	Monolithic	Lerici	Unstiffened
France/Holland/Belgium	Monolithic	Tripartite	Stiffened
France	Sandwich/mono	BAMO	Hull monolithic
Sweden	Sandwich	Landsort	PVC
Norway	Sandwich	Orsoy, alta	SES
Australia	Sandwich	Bay	Catamaran

Aside from MCMV applications a number of other military craft have been built in composites, particularly small patrol craft. A large recent application is the superstructure of the French 'Lafayette' frigates, for which over 80 tons of balsa sandwich are used [18].

4.2 *Offshore structures*

Much recent research and development has been devoted to offshore applications of composite materials. One result of the Piper Alpha enquiry was a less prescriptive approach to materials specification on platforms and increasing use of safety and risk analysis. Composites can be used if they can be shown to result in improved safety. Several recent conferences and papers give a complete overview of this subject (see for example references [19-22]). Significant amounts of composite could be used topside but at present the main application is in the fire protection piping circuits. Other existing and potential applications topside include:

– Walkways, flooring and ladders;
– Tanks and storage vessels;
– Shutdown valve protection, blast panels;
– Accommodation modules.

There has also been some development of high performance composites for subsea components including risers, tethers and drill pipes. Although many applications exist there remains a need for properly validated predictive tools to enable improved design for fire, impact and blast resistance. The designers 'feel' for steel, based on long experience, does not exist yet for composites and any failures which occur assume a high profile.

4.3 *Underwater applications*

High specific compression properties make composites attractive for submersibles and submarine structures which can be subjected to considerable external pressure loadings. The exploration of deep sea will require a tremendous technological effort. Underwater applications are some of the most demanding for materials, and proving reliability is central to the adoption of new concepts. It has been noted that while over a hundred expeditions have reached the world's highest peak only two men have been down to the deepest part of the ocean and that for twenty minutes in 1960 [23]. The key issue in deep sea applications is buoyancy, often expressed in terms of weight/displacement (*W/D*) ratio. A low *W/D* ratio of the pressure hull implies a large positive buoyancy, and the reduction of structural weight enables payload and endurance to be increased as additional exterior buoyancy is not required. Figure 3 shows a schematic illustration of the potential of different materials for deep sea exploration [24].

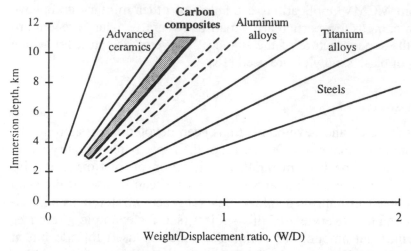

Figure 3. Immersion depths for different materials as a function of *W/D* ratio [24].

Military submarines already use many tons of composites for external decks, domes, and other parts as described by Lemière [25]. Composite material components represent 50% of the wetted area of the 'Triomphant' submarine. Some 1200 m² of composite sandwich, glass/epoxy prepreg skins on a syntactic foam core, are used for the external deck. The use of composites for primary hull structure is currently under study in Europe, USA and Russia. Much of this work is of course confidential but some re-sults from non-military projects are available.

For example, a small unmanned submersible (AUSS) for 6100 m depth was built and tested in the USA recently [26, 27]. This was a wet wound graphite/epoxy cylinder of 0.78 m diameter and 64 mm wall thickness, with titanium end closures. The overall weight to diameter ratio (including end closures) was low, 0.58, showing the potential for these materials, but a number of questions remain to be addressed. The material safety factor for deep sea submersibles are typically around 1.3 for metal alloy hulls, while that for the graphite/epoxy used for the AUSS design was taken as 2. Smith et al have suggested tentative partial safety factors for a trial 6000 m depth composite design [28]:

f_1 (material variability)　　= 1 if lower bound test values used;
f_2 (material degradation)　= 1.20;
f_3 (analysis inaccuracies)　= 1.15;
f_4 (creep, stress rupture)　= 1.15;
f_5 (fatigue)　　　　　　= 1.25.

Thus the overall factor for safety with respect to material failure = $\Sigma f_n \cong 2.0$. Outstanding problems requiring further work can be identified in at least five areas:

1. The development of *manufacturing procedures* for very thick (80 mm or more) filament wound cylinders. This area will benefit from increased experience.

2. The associated *non-destructive control* of such parts, which requires an improved understanding of the effect of defects on the response of composite cylinders to pressure loading so that critical defect sizes can be defined.

3. While buckling failure, which occurs at shallow depths, can be treated to some extent, there is an urgent need for a proven *criterion for material failure* under combined loadings.

4. The generation of cyclic and long term pressure loading data for design.

5. Finally, the *assembly, integration of openings and connections* to metallic components all also require work before manned submarines with composite hulls can be developed.

5 FIRE PERFORMANCE AND REGULATIONS

While there is considerable experience concerning the use of composites many applications are not considered because polymeric composites are believed to show poor resistance to fire. This is not necessarily the case, and the fire behaviour of competing candidate materials should always also be considered.

In general there are two aspects to fire behaviour [29]:
– Reaction to fire; and
– Resistance to fire.

The first covers flammability and the capacity to develop a flame (*M* rating in France, standards NF P 92-501 and 92-507), and toxicity and smoke emission (*F*-rating standard NF F 16-101). The *M* rating is determined using a 500 W epiradiator and a sample of dimensions 400 x 300 mm at 45° to the flame at a distance of 3 cm. The time before inflammation (*Ti*), the sum of the heights of the flames every minute (ΣHm), and the total duration of the flames (δT) are noted. An expression is then used to determine *q* as:

$$q = 100 \times \Sigma Hm\ /\ Ti \times \sqrt{\delta T}$$

This allows classification of materials from *M*1 (non flammable) to *M*4 (easily flammable). The *F* rating involves exposure of a sample to radiant heat using the NBS test and smoke densities are measured over 4 minutes. A tubular oven is then used to burn a 1g sample at 600°C and the gases produced are analysed for CO, CO_2, HCl, HBr, HCN and SO_2 to give a

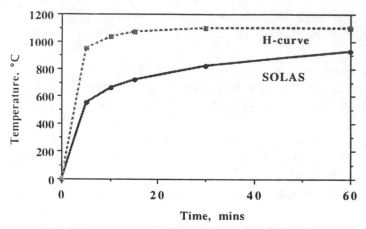

Figure 4. Fire resistance curves for marine applications.

toxicity index. The results from the two tests enable an *F* rating to be given from *F*0 (the best) to *F*5.

Fire resistance tests for marine applications are made using either the SOLAS (IMO resolution A517) heating curve (for shipboard applications) or the Hydrocarbon (or NPD) curve for offshore applications. These are both shown in Figure 4. The H curve is more severe, rising more rapidly and reaching 1100 °C after 60 minutes.

The notion of incombustibility of materials is also often raised in IMO rules, and this is determined by a test used in the building industry, in an oven at 750°C for 20 minutes (ISO 1182). The material must neither produce flames nor cause a significant temperature rise near its surface. Organic matrix composites cannot meet incombustibility requirements and are therefore excluded from many shipboard applications on large vessels. Mineral based materials such as vermiculite or geopolymers are then required. However, there are applications for hulls and bulkheads of smaller ships where an IMO '*F*' classification is acceptable. This requires the temperature rise of the non-exposed surface of a panel subjected to the SOLAS curve to remain below 139°C for 30 minutes. In addition the panel, which has dimensions of 2.44 by 1.91 m, should not allow the passage of flames or smoke during this period. Such tests are expensive but protected sandwich panels can obtain '*F*' ratings.

The H curve has been used in the 'Marinetech' programme [30] to evaluate blast/fire protection panels for offshore applications. Ceramic core sandwich materials could obtain a satisfactory H120 rating if at least 9 mm thick glass/polyester composite skins were used on a 35 mm thick (non-structural) core. Thick composites have impressive fire integrity and a case often quoted is that of a fire in the engine room of HMS Ledbury, which lasted four hours and melted aluminium fittings but the fire was contained by the composite structure of the compartment [31]. The low thermal conductivity of composites (up to 200 times less than steel) is an important factor but the low transport properties of residual glass, the endotherm during decomposition and the cooling effect of volatiles may also contribute to the fire resistance of composites [32].

6 CURRENT AND FUTURE RESEARCH ACTIVITIES

In spite of the extensive list of applications given above there remain a number of other aspects of the application of composites in the marine environment which require further work. For example:

It might be thought that the *ageing of composites* in water now holds few secrets for naval designers but this is not the case. Materials have been immersed for long periods, very complex models of diffusion kinetics have been proposed and residual strengths have been measured, but the link between moisture contents and properties are made only empirically (e.g. [33]) and for few loading types.

In parallel with ageing studies long term mechanical behaviour predictions also need to be addressed. Work on aerospace materials has enabled a methodology to be developed [34] but it has not yet been applied to marine composites. One of the problems encountered is the evolution of polyester properties with time as they are often not completely cured during fabrication [35].

The issue of *fire resistance* is described in section 5. Predictive methods are badly needed but this also entails a requirement for reliable materials data which is only starting to be addressed now.

Composite panels can be designed satisfactorily, and often predicted behaviour is reasonably close to the measured response, at least for quasistatic loading. A recent BRITE/EURAM programme included a number of examples of test-finite element correlation's for structural elements which illustrated this [36, 37]. However, the successful *assembly of composite panels to each other and to metal structures* is still very dependent on experience.

The *impact loading of sandwich composites* has not received the attention given to aeronautical materials, and the response of sandwich materials to wave loading may be the dimensioning loading for fast ships. Weight optimization then becomes important and in order to be competitive with other materials it is essential that existing models be correlated with experimental data. More tests on realistic specimens, such as those performed at DNV on two meter wide V-sections, are required [38].

7 CONCLUSIONS

The marine industry offers enormous possibilities for composite materials. Unique combinations of properties, and retention of these properties in severe environments, will continue to attract the interest of shipbuilders and offshore designers. However, the optimization of materials selection, (which will not always favour the composite material option), can only be performed if the tools for the prediction of structural integrity are available and validated. In too many key areas the designer is forced to rely on empirical safety factors because detailed models are not available. The urgent need for predictive models in the areas of fire, impact response and long term behaviour has been recognized and is currently being addressed.

ACKNOWLEDGEMENT

Contributions to this paper from colleagues at Ifremer, in particular J.-F. Rolin and J. Croquette, are gratefully acknowledged. Results presented in Figure 1 were obtained during the French 'Projet National Composites', co-ordinated by Mme Baudin of the Bureau Veritas and financed by the Ministère de l'Industrie. Fibre content and void contents were determined by Professor Abadie at the Université de Montpellier II.

REFERENCES

[1]   Smith, C.S. Design of Marine Structures in Composite Materials, Elsevier Applied Science, 1990.
[2]   Nautical Construction in Composite materials, *Proc. 3rd Ifremer conference, Dec. 1992*, ed. P. Davies & L. Lemoine.
[3]   *Composite materials in Maritime structures*, ed. Shenoi R.A. & Wellicome J.F., Cambridge Ocean Technology series, 1993.
[4]   Springer, G.S., Ed., *Environmental effects on composite materials*, Vols 1, 2, 3 Technomic Press 1981, 1984, 1988.
[5]   Wolff, E.G., Moisture effects on polymer matrix composites, *SAMPE Journal* 29, 3, May/June 1993, p. 11.
[6]   Gutierrez, J., Le Lay, F. & Hoarau, P., *A study of the aging of glass-fibre resin composites in a marine environment*, in ref. 2, p. 338.
[7]   Hogg, P.J. & Hull, D., Corrosion and Environmental deterioration of GRP, *Developments in GRP Technology 1*, ed. B. Harris, Applied Science Publishers 1983, Chapter 2.
[8]   Jacquemet, R., Etude du comportement au vieillissement sous charge de stratifiés polyester/verre E en milieu marin, Doctorat thesis (in French), IFREMER June 1989.

[9] Farrar, N.R. & Ashbee, K.H.G., Destruction of epoxy resins and of glass-fibre-reinforced epoxy resins by diffused water, *J. Phys. D: Appl. Phys.* Vol. 1978 p. 1009.

[10] Castaing, P., Tsouvalis, N. & Lemoine, L., *Mechanical property degradation of gelcoated glass fiber reinforced polyesters in seawater, in ref. 2, p. 347.*

[11] Graner, W.R., Marine Applications, *Handbook of Composites*, ed. G. Lubin, 1982, Van Nostrand, p. 699.

[12] Baudin, M., *The national project on composite materials in naval technology,* in ref. 2 p. 301.

[13] Gullberg, O., *Non-conventional vessels in Scandinavia made of composite materials,* in ref. 2, p. 27.

[14] Rolin, J.-F., Design and manufacturing of the NES 24 structure, *Proceedings 1st Fast Sea Transportation conference*, FAST 91, June 1991, p. 475.

[15] Rolin, J.-F., *Design and qualification of sandwich composites to be used in the structure of a fast vessel,* in ref. 2 p. 48.

[16] Croquette, J., Parquic, J.C., Forestier, J.M. & Dufour, X., Composite tubes. A *prototype application on a fishing vessel,* in ref 2, p. 412.

[17] Nilsson, J. & Nuss, K.E., Swedish GRP sandwich hull design and shock verification, *Proc. Sandwich Construc*tions 2, March 9-12 1992, Florida.

[18] LeLan, J.Y., Parneix, P. & Gueguen, P.L., *Composite Material Superstructures,* in ref. 2, p. 399.

[19] Gibson, A.G., *Composite structures in offshore applications*, ch. 11 in ref. 3.

[20] *Proc. Conference on Composite Materials in the Petroleum Industry*, IFP 3-4 November 1994.

[21] Robertson, K.A., Troll Phase One Composite usage in a high reliability long life cycle project, *Proc. IIR conference on Applications and Advancements in Composite Materials for the Offshore Industry*, Aberdeen December 1994.

[22] Maritime and Offshore use of Fibre Reinforced Composites, *Proceedings of conference*, Newcastle, June 1992.

[23] Hawkes, G.S. & Ballou, P.J., 'The Ocean Everest concept: A versatile manned submersible for full ocean depth, *Marine Sci & Tech. Jnl.* 24, 2, jun 1990, p. 79.

[24] Girard, D., Les sous-marins d'exploration, *La Recherche 256* Suppl. juillet-août 1993, Vol 24, pp. 844-848.

[25] Stachiw, J.D. & Frame, B., Graphite fiber reinforced plasyic pressure hull mod 2 for the advanced unmanned search system, *NOSC technical report 1245*, August 1988.

[26] Garvey, R.E., Composite hull for full ocean depth, Marine Sci & Tech. Jnl. 24, 2, jun 1990, p. 49.

[27] Lemière, Y., *The evolution of composite materials in submarine structures,* in ref 2, p. 441.

[28] Smith, C.S., Smart, C., Murphy, P. & D.J. Creswell, Design of composite pressure hulls for Autosub vehicles, *ARE report TM 90224*, June 1990

[29] Croquette, J. & Baudin, M., *Commonly-used composite materials: Fire behaviour,* in ref. 2, p. 287.

[30] Worrall, C.M. & Gibson, A.G., *Design of panels for fire resistance,* in ref. 22.

[31] Nixon, J.A., *Ship experience and panel design,* in ref. 22.

[32] Gibson, A.G., Wu, Y.-S., Chandler, H.W., Wilcox, J.A.D. & Bettess, P., *A model for the thermal performance of thick composite laminates in hydrocarbon fires,* in ref. 20.

[33] Pritchard, G. & Speake, S.D., The use of water absorption data to predict laminate property changes, *Composites*, *18*, 3, July 1987, p. 227.

[34] Dillard, D.A., *Viscoelastic behavior of laminated composite materials, Fatigue of composite materials*, ed. K.L. Reifsnider, Elsevier 1990, Ch 8.

[35] Cardon, A., De Wilde, W.P. & De Munck, L., Viscoelastic characterization of glass fiber/epoxy and glass fiber/polyester composite materials, Vrije Universiteit Brussel report Ifremer-VUB contract 1992.

[36] Davies, P., Choqueuse, D. & Bigourdan, B., Test-finite element correlations for non-woven fibre reinforced composites and sandwich panels, *Marine Structures*, 7, 1994, p. 345.

[37] Composite materials in marine structures and components, Final report of 'COMAST' BRITE/EURAM project BREU 0178-C, October 1993.

[38] Hayman, B., Haug, T. & Valsgaard, S., Slamming drop tests on a GRP sandwich hull model, *Proc. 2nd conf. on Sandwich construction*, Florida 1992.

Fatigue behaviour of polymer-based composites and life prediction methods

Bryan Harris

School of Materials Science, University of Bath, Bath, UK

ABSTRACT: A general introduction to the problems of defining and studying the fatigue behaviour of composite materials is presented, with specific reference to the high-performance carbon-, glass-, and aramid-fibre-reinforced thermosets that are of current interest to the aerospace industry. A description is also given of the overall patterns of fatigue response of composites, with particular emphasis on materials of unidirectional and $[(\pm45,0_2)_2]_S$ lay-up, with reference to plain and hybrid systems, and including consideration of damage accumulation and the effects of damage on residual properties. A method of analysis of data in terms of constant-life diagrams for life prediction purposes is presented, together with an indication of the level of success that may be achieved in prediction and prevention of the potential hazards. Some consideration is also given to statistical aspects of data analysis. Mention will be made of the problem of variable-stress testing and the difficulties of analysis of the data from such tests and the paper concludes with a brief discussion of the use of artificial neural networks for life prediction purposes.

1 INTRODUCTION

Some years ago, it was not uncommon for users of composite materials, even within the aerospace industry, to express the belief that composite materials – specifically carbon-fibre-reinforced plastics – did not suffer from fatigue. This was an astonishing assertion, given that from the earliest days of the development of composites, their fatigue behaviour has been a subject of serious study. What it usually meant was that because most CFRP were extremely stiff in the fibre direction, the working strains in practical components at conventional design stress levels were usually far too low to initiate any of the familiar local damage mechanisms that might otherwise have led to deterioration under cyclic loads. Fatigue failures did

49

not therefore appear to occur – hence the above assertion – although it is certain that, had it been sought, evidence of fatigue damage would certainly have been found.

The idea of using composites, and especially CFRP, only at very low working strains begs two important questions. The first is the obvious one that by using expensive, high-performance materials at small fractions of their real strength, we are over-designing or, in more cost-conscious terms, we are not using them economically. The second is that since anisotropy is a characteristic that we accept and even design for in composites, a stress system that develops only a small working strain in the main fibre direction may easily cause strain levels normal to the fibres or at the fibre/resin inter- face that will be sufficiently high to cause the kind of deterioration that we call fatigue damage. In designing appropriately with composites, therefore, we cannot ignore the fact that fatigue is a factor that must be taken into consideration. And it follows directly that, in addition to needing to under- stand the mechanisms by which damage occurs in composites, we need ac- cess to procedures by which the development and accumulation of this damage, and therefore the likely life of the material (or component) in question, can be reliably predicted.

2 REVIEW OF SOME CLASSICAL FATIGUE PHENOMENA

2.1 *Stress/life curves*

A good deal of the work that has been done on the investigation of the fa- tigue of fibre composites reflects the much more extensive body of knowl- edge relating to the fatigue of metallic materials. And this is not altogether unreasonable since the established methods of accumulating and analysing metallic fatigue data provided a reliable means of describing fatigue phe- nomena and designing against fatigue. The formidable treatise of Weibull (1959), for example, remains a valuable source book for modern workers on composite fatigue. The danger was, and is, in making the assumption that the underlying mechanisms of material behaviour that give rise to the stress/life (σ/log N_f) curve are the same in metals and composites.

Before the development of fracture mechanics and its use in treating metallic fatigue as a crack-growth problem, the only available design in- formation on fatigue behaviour was the stress/life, or *S/N* curve (Fig. 1). It represented directly the perceived nature of fatigue in terms of experimental results, but gave no indication of the mechanisms of fatigue damage, of the presence or behaviour of cracks, or of changes in the characteristics of the material as a consequence of the fatigue process. The curve represents the

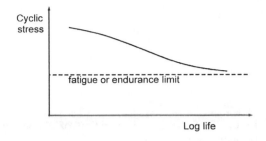

Figure 1. Schematic illustration of a metallic *S/N* or σ/log N_f curve.

stress to cause failure in a given number of cycles, usually either the mean or median life of a series of replicate tests at the same stress. Despite the anomaly (from a mathematical viewpoint) of plotting the dependent variable on the abscissa rather than vice versa, the *S/N* curve is nevertheless a useful starting point for the designer, provided (and only provided) due attention has been paid to the statistical aspects of data generation, so that the apparently simply failure envelope which it defines is associated with failure probabilities rather than with some simplistic fail/no-fail criterion, i.e. it is presented as an *S/P/N* curve. It can then be used, as many designers prefer, without any consideration of the underlying fatigue damage mechanisms that will lead to failure, despite the availability of rather better fracture-mechanics methods which do.

There is logic in using the semi-logarithmic form of the σ/log N_f plot but, despite Basquin's Law, no fundamental reason for supposing a priori that the failure envelope would be linear on a log-log plot. Some stress/life plots are linear on both kinds of plot, some on neither. The knowledge that for bcc metals and alloys, like steels, for example, the curve flattened out at long lives (> 10^7 cycles, say) and gave rise to a fatigue or endurance limit, σ_e, which could be explained in terms of dislocation/solute-atom models, and that the fatigue ratio, σ_e/σ_o, where σ_o is the monotonic tensile strength, was roughly constant for a given class of alloys, gave confidence to users of this kind of design data. Finally, there was no general idea that the metallic σ/log N_f curve should necessarily extrapolate back to a stress level (at $N_f = 1$ cycle, for example) which was simply related to any monotonic strength property of the material.

2.2 *Constant-life diagrams*

In order to use stress/life information for design purposes, a common method was to cross-plot the data to show the expected life (or expected life at some particular probability level) for a given combination of the alternating component of stress, σ_{alt}, defined as ½ $(\sigma_{max} - \sigma_{min})$, and the mean stress, σ_m, which is ½ $(\sigma_{max} + \sigma_{min})$. This automatically introduces the con-

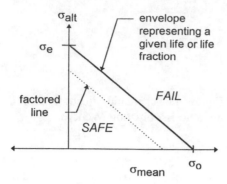

Figure 2. Schematic illustration of a constant-life or Goodman diagram.

cept of the stress ratio, $R = \sigma_{min}/\sigma_{max}$, and the question of the relative importance of any compression component of stress. In metallic fatigue it was frequently assumed, however, that compression stresses were of no significance because they acted only to close fatigue cracks, unlike tensile forces. Master diagrams of this kind were presented in a variety of forms, all more or less equivalent, but the most familiar is that which is usually referred as the Goodman diagram (Fig. 2). For design purposes, it was useful to have an equation to represent the fail/safe boundary in this diagram, and it is of course the linear relationship that is associated with the name of Goodman (1899) although others have been proposed, including the parabolic relationship of Gerber (1874), and the linear and parabolic 'laws' have been modified to include safety factors on one or both of the stress components. An important question, for metals as well as for composites, relates to the minimum amount of test data that is needed to define the failure envelope with a level of reliability which is sufficient for engineering design of critical components.

2.3 *Variable-stress fatigue and cumulative damage laws*

We have so far been considering the use of data from fatigue experiments in which the stress level or stress range remains fixed during the test, although it must be admitted that for metallic samples displaying fatigue crack growth the concept of a constant stress is dubious. A satisfactory life-prediction method, however, must be able to cope with variable stress.

In most fatigue environments the mean and alternating stress amplitudes vary and may be presented as a spectrum of the frequency of occurrence of different stress levels. The concept of deterioration as a result of damage accumulation within the material with stress cycling is used, linked to the notion of residual strength, and the damage is quantified such that a parameter d represents the fraction of catastrophic damage sustained after n cycles ($n < N_f$). At failure, $d = 1$ and $n = N_f$. Thus, d_i represents the frac-

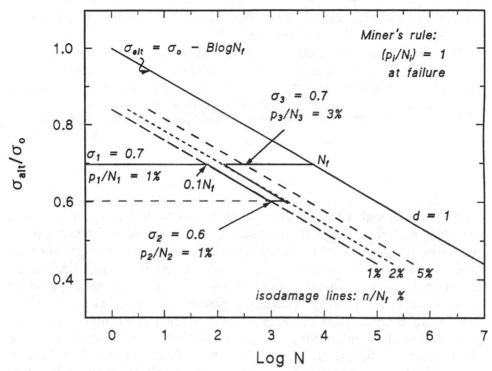

Figure 3. Schematic illustration of the application of the linear damage law and the concept of iso-damage lines.

tional damage after cycling at a stress σ_i for a fraction of life n_i/N_i. Various proposals for damage laws have been derived, the most common of which is the Palmgren-Miner (Miner, 1945) linear damage rule. The accumulation of damage is supposed linear with number of cycles, and is apparently independent of the value of the stress:

$$d = n/N \qquad (1)$$

The application of the simple damage model is illustrated schematically in Figure 3. If a stress, σ_1, which is 70% of the failure stress, σ_o, is applied for 1% of the normal life, N_1, of the sample at that fatigue stress, it may be considered to have reached a notional curve, as illustrated, which represents a state of damage corresponding at all stress levels to 1% of the damage level associated with actual failure, however that damage level might be assessed. If the stress level is then reduced to σ_2, which is only 60% of the sample strength, and cycled for a further 1% of its life at the new stress, we must first follow the path shown down the initial 1% damage line to a position representing an iso-damage state vis-à-vis the life N_2. After a further 1% of damage at the lower stress, we imagine the initial stress level, σ_1, to be resumed for a further 3%, when the overall damage state of the sample now represents 5% of failure damage.

Thus, if the load spectrum is divided into blocks with Δn_i (which we now define as p_i) cycles at stress levels σ_i, then Equation 1 applies incrementally to each block and the total damage is the sum of the individual 'block' damages, irrespective of the order of their application, so that:

$$\sum \Delta d = \frac{\sum \Delta n_i}{N_i} \ (= 1 \text{ for failure}) \tag{2}$$

This well-known, though scarcely ever obeyed, rule was originally proposed for the prediction of the life of metallic components undergoing fatigue, and although a stock feature of text-book presentations of fatigue it has always been viewed with great suspicion by designers because it is often found to give non-conservative results, i.e. to predict lives greater than those observed experimentally. The simple concept of damage defined by Equation 1 is not, of course, a mechanistic concept. It has been applied to many kinds of engineering materials, regardless of the actual nature of the damage mechanisms that contribute to the gradual deterioration and ultimate failure of an engineering component subject to cyclic loading. In the field of metallic fatigue, the logical definition of damage must be related to crack growth, and the need for an empirical cumulative damage law was perhaps obviated by the arrival of fracture mechanics. In materials where fatigue deterioration cannot be easily described in terms of single crack growth – composite materials, ceramics and concrete are typical examples – effort continues to be expended in searching for damage parameters that may be used in life prediction.

The simplest step forward from the linear damage rule is to look for non-linear functions that still employ the damage parameter d, as defined by Equation 1. In the Marco-Starkey (1954) model, for example, a simple non-linear presentation suggests an Equation of the form:

$$d = (n/N)^\alpha \tag{3}$$

where α is a function of stress amplitude. The condition for failure when $d = 1.0$ is satisfied and the Miner rule is a special case of this law with $\alpha = 1.0$. As shown in Figure 4, in the linear case the line represents all stress levels and increments of damage equate to increments of fractional life, whereas in the non-linear case each stress level requires a different curve and equal increments of life fraction do not produce equal increments of damage fraction. Thus, we need to distinguish between the life fraction, n/N_f, where n is measured from zero, as in Equation 3, and the incremental life fraction, p/N_f, applied in a block-loading programme.

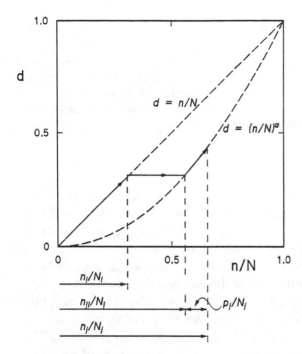

Figure 4. Schematic illustration of linear and non-linear damage laws.

3 FATIGUE PHENOMENA IN REINFORCED PLASTICS

3.1 *Tensile stress/life curves*

Since the monotonic loading of a fibre composite induces a variety of microstructural damage mechanisms, including fibre failures, matrix cracking, interfacial debonding and so forth, it is self-evident that repeated loading is likely to continue developing this damage, even when the overall stress level is well below the normal failure stress. By contrast with the nature of 'fatigue damage' in metallic materials, then, which relates to crack growth rather than to any change in the actual mechanical properties of the material, it is apparent that the progressive accumulation of local damage in reinforced plastics could, in due course, lead to a deterioration in the strength and stiffness of the material. And if the damage continues until the residual strength of the composite falls to about same level as the externally applied load, the material will fail. This sequence of events will of course give rise to an $\sigma/\log N_f$ curve that will resemble in many ways those for metallic materials, although the underlying reason for the phenomenon in fibre composites is quite different. The process, which is often referred as 'wear-out', is illustrated in Figure 5.

This rather simple picture disguises the fact that the dominant damage

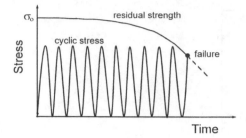

Figure 5. Degradation of composite strength by wear-out strength falls from the normal composite strength, σ_o, to a level where the residual strength equals the magnitude of the fatigue stress, σ_f, at which point failure occurs.

mechanisms vary both with the nature of the composite (the particular combination of fibres and matrices, reinforcement lay-up, etc.), and with the loading conditions (tension, bending, compression, etc.), and results for any one material under a given stress condition may appear to fit no general pattern. This is easily illustrated with reference to the gradual accumulation of results and materials development over the last two decades or so.

First, the forms of the stress/life curve for various kinds of glass-reinforced plastic (GRP) have long been familiar to researchers, as was the fact that the fall of the curve with reducing stress was considerable. This fall, or average slope, is an indication of the fatigue resistance of the material, a more rapid fall being indicative of a greater susceptibility to fatigue damage. Mandell et al. (1982) have attempted to show that fatigue damage in a many GRP materials can be explained simply in terms of gradual deterioration of the load-bearing fibres. They analyse a great range of experimental data to demonstrate that the behaviour of composites with long or short fibres, in any orientation, and with any matrix, can be thus explained. They do this by effectively forcing $\sigma/\log.N_f$ curves to fit a linear law of the form:

$$\sigma = \sigma_f - B \log N_f \tag{4}$$

where σ is the peak cyclic (tensile stress), σ_f is the monotonic tensile fracture stress, N_f, is the fatigue life of the material at σ_f, and B is a constant. For a wide and disparate range of GRP materials, including moulded reinforced thermoplastics, they found that the stress ratio σ_f/B had a constant value of about 10 with very little dispersion. In principle, the idea that the controlling mechanism in fatigue of composites is the gradual deterioration of the load-bearing fibres is logical and inevitable. What controls the actual life of any given sample or composite type is simply the manner in which other mechanisms, such as transverse ply cracking in 0/90 laminates, or local resin cracking in woven-cloth composites, modify the rate of accumulation of damage in the load-bearing fibres. Nevertheless, a single mechanistic model which includes randomly-reinforced, dough-moulding com-

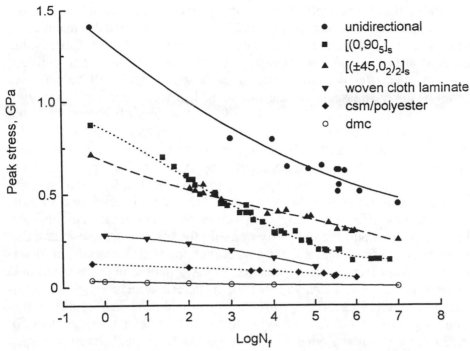

Figure 6. Repeated tension stress/life curves for various types of GRP materials ($R = 0.1$) (Harris, 1994).

pounds and injection-moulded thermoplasts as well as woven and non-woven laminates is unexpected.

One of the difficulties is that the $\sigma/\log N_f$ curves for many GRP materials are not linear, as shown by the selection of data for a group of glass-fibre composites shown in Figure 6 (Harris, 1994). It can be seen from this selection of results that the curve for one of the materials, a 0/90 cross-plied laminate, flattens out at low stresses to suggest what, in metallic fatigue, we have referred to as an endurance limit. With GRP there are two factors which complicate the appearance of the stress/life curves. As a consequence of the high failure strain of the glass fibres and their sensitivity to moisture, a) the tensile strengths of GRP materials are strain-rate and temperature sensitive, and b) during cycling at large strains there is usually a significant rise in temperature as a result of hysteretic heating which is not easily dissipated by the non-conducting constituents of the GRP. The consequences of this have been demonstrated by Sims & Gladman (1978). First, when fatigue tests are carried out at constant frequency over the whole stress range of interest, the deformation rate is usually considerably greater than the rates normally used for measuring the monotonic strength, and as a result, since it is common for the strength to be included, as shown in Figure 6, as the point on the extreme left of the $\sigma/\log N_f$ curve (i.e. notionally at $N_f = 0.5$ for a repeated tension test, or $N_f = 0.25$ for a fully reversed tension/compression test) that measured strength is often actually lower than a value measured at the fatigue-test frequency. Second, if tests are run at constant frequency, the effective strain rate at each stress level will be different (higher at lower stresses), and the measured life values for each stress will not be associated with a common base-line of material behaviour: i.e. the fatigue curve will refer to a range of stresses σ which are proportions of a variable material property, say $\sigma_f (\dot{\varepsilon})$, instead of a true material property that we normally define as the strength, σ_f. And third, as the peak stress level falls and the life of the sample extends, the level of hysteretic heating rises, and the effective baseline strength of the material therefore also falls, so reversing the effect of the higher effective deformation rate. As Sims and Gladman pointed out, the only way to avoid the effects of these interacting processes is to ensure that all $\sigma/\log N_f$ curves for GRP materials are determined at a fixed rate of load application (RLA), that the material strength is measured at the same RLA, and that the hysteretic effects are either eliminated or accounted for. These corrections can make substantial differences to the shapes of stress/life curves and to the validity of the design data obtained from them: many otherwise curved $\sigma/\log N_f$ graphs are also rendered more linear by these corrective measures.

By contrast with GRP, carbon-fibre composites are largely rate insensitive and because they deform less than GRP under working loads and are reasonably highly conducting, hysteretic heating effects are usually insignificant (Jones et al., 1984). Some of the earliest CFRP, reinforced with the HM (high modulus, or Type 1) species of carbon fibre had such low failure strains that fatigue experiments in tension were extremely difficult to carry out, and some of the first tension $\sigma/\log N_f$ curves published (Beaumont & Harris, 1972; Owen & Morris, 1972) were almost horizontal. The S/N line hardly fell outside the scatterband of measurements of the monotonic tensile strength, the fatigue response being effectively dominated by tensile failure of the fibres alone and the slight reduction in resistance after 10^6 or 10^7 cycles appeared to be a result of slight viscoelastic (creep) strains rather than to cycling effects per se (Fuwa et al., 1975). Only in torsion or bending tests was the slope of the $\sigma/\log N_f$ increased as the changed stress systems allowed other damage mechanisms to occur.

A very small failure strain is often a disadvantage in engineering materials (some of the early CFRP were extremely brittle) and developments in carbon-fibre manufacture led to materials that possessed higher failure strains (lower elastic moduli), although these improvements were often offset by concomitant reductions in strength. By the end of the 1970s, the most common polyacrylonitrile- (PAN) based fibres, designated HMS, HTS and

XAS, covered a range of moduli from about 400 GPa down to 200 GPa. It had been shown (Sturgeon, 1975) that the tensile stress/loglife data for these materials fell approximately on straight lines with slopes which increased with decreasing fibre stiffness. Taking Sturgeon's results together with their own data for unidirectional and 0/90 lay-ups of both HTS carbon and E-glass fibres, Jones et al. (1984) showed that the slope of the $\sigma/\log N_f$ curve (B in Equation 4) normalised with respect to the composite strength (B/σ_f) was an inverse linear function of the reinforcing fibre modulus. The effect of the normalisation with respect to strength effectively eliminates variations due to differences in fibre volume fraction between composites. Since this simple strain-controlled model applied equally well to unidirectional and 0/90 laminates of the same composite, it also suggested that fatigue response was substantially controlled by the axial load-bearing fibres and that the easy cracking in the transverse plies did little to impair the fatigue resistance of the $0°$ plies. Despite the fact that the $\sigma/\log N_f$ data for some of the composites involved in this analysis still had to be forced to fit a linear curve, the analysis is in accord to some extent with the Mandell model with the difference that Mandell's correlation was with composite strength, whereas that of Jones et al. was with fibre stiffness (or, indirectly, composite strain).

Increasing awareness on the part of designers of the qualities of fibre composites, and of carbon-fibre composites in particular, resulted in demands for fibres with combinations of strength and stiffness which were different from those which characterised the earlier materials. A particular call was for higher failure strains in association with high strength, and these demands led to developments in carbon-fibre manufacture and the availability of a much wider range of fibre characteristics. The fatigue performance of composites based on these newer fibres is clearly different from that of earlier CFRP, as illustrated by the range of results shown in Figure 7. These data are all for composites with the lay-up $[(\pm 45,0_2)_2]_s$ and the curves are all polynomial fits to experimental data obtained at the University of Bath. The results for the composites based on the early Courtauld XAS fibres are similar at the higher stress levels, but the separation of the two curves labelled 1 and 2 at lower stresses is presumably due to the effect of the different matrix resins, 913 and 914. The data points for these two curves actually overlap to a large extent, and we had previously considered the two composites as having identical fatigue responses, i.e. a common linear $\sigma/\log N_f$ curve (note the stress axis scale in Fig. 7). The similarity was even more obvious in unidirectional samples of the same two materials (Harris et al., 1990). Curve 3 represents a composite containing a more recent fibre, the Enka HTA material, in the older 913 epoxy resin. This fibre, similar to Toray T300 and in principle little different from XAS, neverthe-

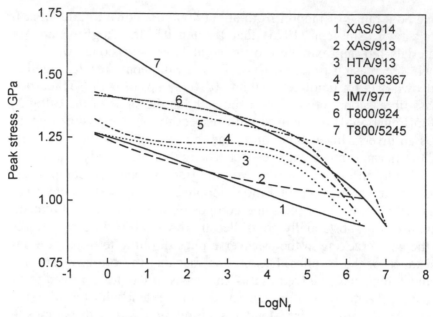

Figure 7. Median σ/log N_f curves at $R = 0.1$ for seven varieties of carbon-fibre composite with the $[(\pm45,0_2)_2]_s$ lay-up.

less generates a completely different form of σ/log N_f curve which shows an initially low slope (high fatigue resistance, followed by a rapid fall to the level of the curve for XAS/914. This form of curve is then almost exactly reproduced (curve 4) for the T800/6367 composite, surprising, since the T800 fibre is a much stronger fibre. The importance of the resin is clearly shown in this diagram, since it includes three composites based on T800 with different matrices. The T800/5245 (the resin a bismaleimide epoxy) is the composite in which the superior properties of the T800 fibre are made best transferred into the composite and this superiority is also reflected in the shorter-term fatigue properties. However, the superiority is rapidly lost and at the longest lives, the fatigue resistance is actually worse than that of the humblest of these composites, the XAS/913. Perhaps the most notable feature of this collection of data is that whereas the earliest of the composites had approximately linear, or even slightly upward turning σ/log N_f curves like most GRP materials, the newer, increasingly 'high-performance' composites show downward curvature. We note that none of these curves for the newer materials have reached any kind of stable fatigue limit within the convenient laboratory test window of extending to 10^7 cycles.

This downward curvature has also been observed in composites reinforced with Kevlar-49 aromatic polyamide (aramid) fibres (Jones et al., 1984; Fernando et al., 1988). Typical behaviour of a KFRP composite is

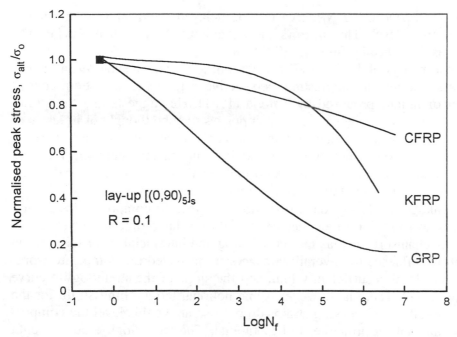

Figure 8. Comparison of stress/life curves for three comparable fibre composites with 0/90 lay-up and a common Code-69 epoxy resin matrix.

compared with those of similar GRP and CFRP materials in Figure 8. The composites are comparable in every respect except for the reinforcing fibres (the carbon is HTS fibre). Two aspects of the behaviour of the Kevlar composite deserve comment. The performance of KFRP over the first three decades (i.e. high stress levels) is superior even to that of the CFRP on a normalised basis (although the KFRP composite had a somewhat lower tensile strength. At maximum cyclic stress levels only marginally below the tensile strength scatter band, however, the fatigue resistance deteriorates rapidly, and a change in controlling failure mode is clearly indicated. At the highest stress levels failure occurs by the normal fibre-dominated failure process. But since the normalised slope of this high-stress region is higher than that of most of the CFRP laminates in Figure 7 it is clear that the higher toughness or resilience of the organic fibres, which results from their complex structure and failure mode (Dobb et al., 1979; Bunsell, 1975), leads to a high fatigue resistance. At lower stresses, however, when the samples are run for much larger numbers of cycles, the internal structural weaknesses of the Kevlar-49 filaments are exposed to the effects of the cyclic stress and the fibres themselves exhibit much lower fatigue resistance. This apparent weakness can be overcome to some extent by preventing complete unloading of the composite between cycles. Effectively, the downturn of the $\sigma/\log N_f$ curve can be delayed by raising the minimum cyclic stress level. This mirrors the behaviour of individual Kevlar filaments as described by Bunsell (1975).

These results of Jones et al. for carbon, glass and Kevlar-49 laminates may be put into an interesting perspective relative to the strain-control model of fatigue proposed by Talreja (1981). He suggests that the strain/loglife curves for polymer matrix composites may be thought of in terms of three régimes within which separate mechanisms control fatigue failure. At high stress levels fibre breakage (and/or interfacial debonding) occurs which leads to failures within the normal tensile failure scatter band of the 0° plies. At somewhat lower cyclic stress levels, however, this statistical fibre breakage, while still occurring, does not lead to composite destruction in lives so short that other mechanisms do not have time to occur. These other mechanisms. such as matrix cracking and interfacial shear failure, can therefore influence the overall damage state provided the composite working strain level is sufficiently high and the slope of the strain/loglife curve begins to fall. There is, however, some notional fatigue limit strain for the matrix itself and if working strains do not rise above this level the composite should not, in principle, fail in fatigue. The strain/loglife curve ought therefore to flatten out again and something in the nature of an endurance limit should be observed. Whether or not some or all of these stages are observed will clearly depend on the characteristics of the constituents and the lay-up geometry.

The model is well illustrated by replotting the data of Figure 8 in terms of maximum cyclic strain versus life, as shown in Figure 9, where the influence of the fibre deformation characteristics is clearly visible. High working strains in the GRP prevent the establishment of Stage 1 of the Talreja model, the curve moving almost immediately into Stage 2, whereas the higher modulus of the Kevlar fibre delays this transition in the KFRP composites. Working strains in the CFRP are rarely sufficiently high as to exceed the matrix fatigue limit and so the strain/life curve (like the $\sigma/\log N_f$ curve) retains the low slope, characteristic of Stage 1, for the entire stress range shown. It is apparent from the shapes of the selection of $\sigma/\log N_f$ curves for various materials shown in the last four figures, that many of these high-performance composites exhibit similar features in tensile fatigue, regardless of the specific nature of the degradation mechanisms, although the extent to which some or all of the stages of Talreja's model are visible within the practical experimental testing window depends on the nature of the constituents and of the fibre/matrix bond.

This discussion has so far been concerned only with repeated tension fatigue: consideration will be given to the effects of compression in the later discussion of constant-life models.

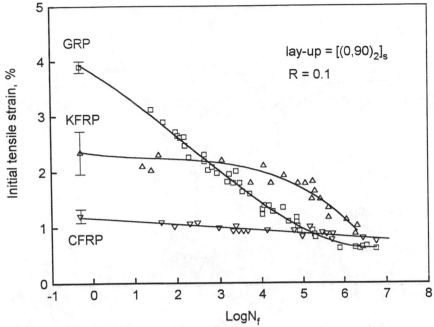

Figure 9. Strain/life curves for $[(0,90)_2]_s$ laminates of composites with HTS-carbon, Kevlar and E-glass fibres in a common epoxy resin (Code-69).

3.2 *Residual strength*

We referred earlier to the wear-out model of fatigue damage. It is a matter of common experience that both the rigidities and strengths of composite materials are affected adversely by repeated loading as damage accumulates in the material (see, for example, Reifsnider & Jamison, 1982; Adam et al., 1986; Chen & Harris, 1993). In some circumstances, both strength and stiffness may initially increase slightly as the fibres in off-axis plies or slightly misaligned fibres in 0° plies re-orient themselves in the visco-elastic matrix under the influence of tensile loads, but the dominating effect is that due to the accumulating damage in the material. The deterioration during cycling usually depends on the stress level, as shown in Figure 10.

In their work on the comparable group of CFRP, GRP and KFRP 0/90 laminates already referred to, Adam et al. (1986) observed that the shapes of the residual strength curves for all three materials were similar and could be well represented by an interaction curve of the form:

$$t^x + r^y = 1 \tag{5}$$

where t is a function of the number of cycles, n, sustained at a given stress, σ_{max}, for which the expected fatigue life is N_f.

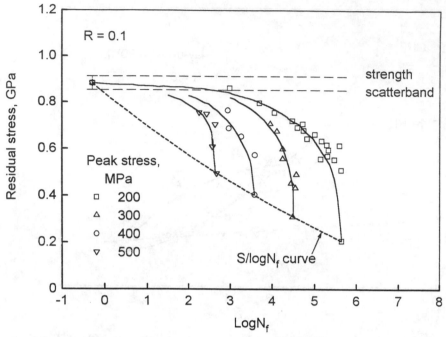

Figure 10. Residual strengths of samples of a 0/90 GRP laminate after tensile fatigue cycling at various stresses.

$$t = \frac{\log n - \log 0.5}{\log N_f - \log 0.5} \tag{6}$$

The factor 0.5 simply accounts for the fact the normal strength of the material corresponds to the lower limit of the cycles scale at ½ cycle. The normalised residual strength ratio, r, is defined as:

$$r = \frac{\sigma_R - \sigma_{max}}{\sigma_f - \sigma_{max}} \tag{7}$$

The exponents x and y are materials-dependent parameters which may have physical significance although for the original paper they were obtained by curve fitting. The normal monotonic tensile strength, σ_f, and the residual strength after cycling, σ_R, are determined at the same loading rate as that used for the fatigue cycling. The loss in strength due to cycling, $(\sigma_f - \sigma_R)$, may be regarded as a damage function:

$$d = 1 - r = (1 - t^x)^{1/y} \tag{8}$$

The shape of the curve can vary from a linear form (with $x = y = 1$), through a circular quadrant for powers of two, to an extremely angular

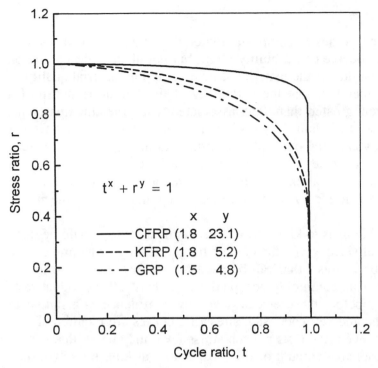

Figure 11. Normalised residual strength curves for three 0/90 composite laminates tested in repeated tension at $R = 0.1$.

variation for very high powers. Since both the gradual wear-out and sudden-death type of behaviour can both be accommodated within a single general model, a treatment of this kind has advantages over non-normalising procedures. Despite apparent dissimilarities in the fatigue behaviour of the three types of composite studied in this work, the residual strength results suggested that a common mechanism of damage accumulation led to final failure since the damage law of Equation 5, with suitably chosen values of x and y, fit all of the data for all three materials, as shown in Figure 11. Having established the values of x and y for a given material, the residual strength can then be evaluated from the equation:

$$\sigma_R = \left(\sigma_f - \sigma_{max}\right)\left(1 - t^x\right)^{1/y} \tag{9}$$

It seems unlikely that a single mathematical model could ever cope with the whole range of composite types and lay-ups available, and the model would require validation for each system of interest – materials, lay-ups, stress systems. But It can be seen that validation would itself be reasonably economical of test time and material, given that the interaction curve could be deduced from relatively few data.

3.3 *Statistical aspects*

The mechanical properties of composite materials are almost always some-what variable, the degree of variability often, but not always, depending on the degree of variability of the material structure, i.e. the material quality. It is not surprising therefore that the variability of the fatigue response of a composite is even greater than that associated with metallic materials. Stress/life data may be obtained by testing single samples at many different stress levels, or by carrying out replicate tests at rather fewer stress levels: the latter is usually considered to be the most satisfactory method because it provides statistical information at each stress level, and permits the drawing of probability/stress/life curves in addition to median life or mean life curves.

One of the problems is to know how many replicate tests should be done at each stress level since, given the cost of fatigue-testing programmes, the smaller the number of tests that can be used to establish a 'safe' σ/log N_f curve, the better. It is commonly accepted that at least 20 individual tests may be necessary before the user can have any confidence in a statistical analysis of results, and yet when, say, stress/life curves are required at five different R ratios, even five tests at each stress level may be all that can be provided in a reasonable amount of time, especially at long lives. Whitney (1981) has suggested that where small numbers of stress/life values are available at a number of different stress levels, the data may be pooled to give an overall value of the Weibull shape parameter, m, this value then being used to obtain working stress/loglife curves for any given failure probability. This is done by normalising each test stress data set with re-spect to the characteristic life, N_o (obtained as the scale factor of the Weibull distribution for the data set), pooling all data sets for all stress lev-els and R ratios, and then re-ranking them in order to allot a new failure probability function to each point. The virtue of the procedure is that a very large population may be used to derive the value of the Weibull shape pa-rameter and that calculations of expected life based on that m value will be highly reliable. The weak point is perhaps the question of the validity of the original assumption that a small group of data can be fitted to a two-parameter Weibull plot and a valid scale factor extracted. Nevertheless, Whitney successfully analysed some data of Ryder & Walker (1977) and showed that the pooled data gave a value of 1.1 for the shape factor. Gathercole et al. recently had similar success in analysing the behaviour of a modern T800/5245 carbon-fibre/bismaleimide-epoxy composite. The best-fit line to their pooled data also gave a Weibull modulus $m = 1.1$, with correlation coefficient 0.98 for 168 data points. Their result was also later confirmed by Lamela-Rey (1994) in a more rigorous analysis based on ex-treme-value theory by Castillo et al. (1993).

It is interesting to note that for an *m* value of unity, which is very nearly the case for these pooled data from the two cases mentioned above, the Weibull model reduces to a simple exponential distribution. This distribution governs systems where failure is independent of life (Chatfield, 1970), i.e. where the probability of failure within a given time interval, Δt, after a service period *t* is independent of *t*, i.e. failure is a random event, and the system does not deteriorate as a result of service. This is not what is normally considered to be so in the case of the fatigue failure of reinforced plastics for which it is known that residual performance is in fact reduced as a result of the accumulation of damage. One reason why the data may appear to fit an exponential distribution is that the range of lives at a given stress level is often very wide, spreading sometimes over two decades or more. And whereas on a logarithmic life axis this may not appear too serious, when considered in the Weibull analysis as a linear function, N_f, rather than as a logarithmic function, log N_f, the range may easily be judged to be bounded at the lower extreme at $N_f = 0$, and hence the distribution may appear to be exponential. In reality, the shape of the Weibull distribution changes from being of classical exponential form to being peaked over a very small range of values of *m*, as illustrated in Figure 12.

In some recent work at Bath University, we have extended this type to

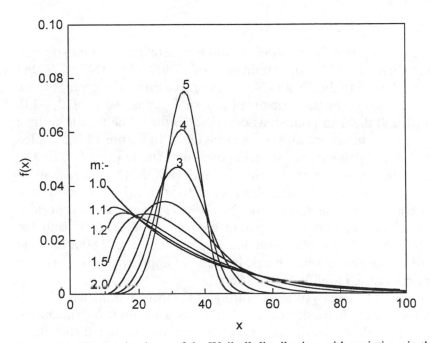

Figure 12. Change in shape of the Weibull distribution with variations in the shape parameter, *m,* near to unity.

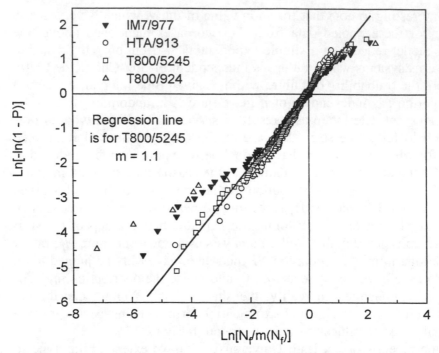

Figure 13.Two-parameter Weibull plots of pooled, normalised fatigue lives for four CFRP laminates.

analysis to cover a group four species of modern aerospace composite materials, including $[(\pm 45,0_2)_2]_s$ laminates of T800/924, IM7/977 and HTA/913, in addition to the T800/5245 mentioned earlier. For each of these materials, $\sigma/\log N_f$ curves were obtained at five R ratios, +0.1, –0.3, –1.0, –1.5 and +10, and the data pooled, as described earlier. The tensile fatigue curves for these laminates are all shown in Figure 7. In Figure 13 the pooled results for all four laminates are superimposed in a single plot. The line in this figure is the linear regression line for the T800/5245 data set, mentioned earlier. It can be seen that the results for the IM7/977 composite show a pattern very similar to that for the T800/924 material and the four data sets therefore fall into two separate groups, the T800/5245 and the HTA/913, which both present good linear plots, and the T800/924 and IM7/977, which show the same break in slope at approximately the same value of the normalised loglife, $\log (N_f/N_o)$.

No extension of this analysis to the fitting of the three-parameter Weibull model has been made here, despite the logic that a location parameter (or minimum life) would obviously be expected for a fatigue-life distribution, because the curvature of the plots for the IM7/977 and T800/924 composites is of the wrong sign to allow them to be linearised by introducing the third parameter.

It is interesting to note that the tail for the IM7/977 data set coincides approximately with that for the earlier T800/924 laminate, and in both cases it comprises only about ten data pairs, roughly the bottom decile of the distribution (the pooled data set for the T800/924 composite comprised 86 fatigue life results, while that for the IM7/977 material contained 105 results). The remaining data sets (indeed, the vast majority of all fatigue lives) for all four laminates clearly form part of a larger population which is satisfactorily fitted by the two-parameter Weibull model with a slope of the order of unity. There is a temptation to accept this as being an indication that the tails of the two non-linear distributions are in some way due to 'poor' data values and, since the tail is at the bottom end of the distribution, that these 'poor' results are actually premature failures, perhaps caused by some initial instability in the testing machine. To treat these points as 'outliers' and ignore them, however, would at this stage be quite inadmissible since it is clearly the minimum-life data that have the greater significance for the designer, rather than the median lives which are often used as the most easily accessible life parameter.

It is not apparent why two of the pooled life distributions exhibit tails indicating poorer reliability than would have been predicted from the Weibull parameters obtained from the remaining linear distributions. Since the use of the pooling approach is intended to predict safe lives from limited data sets, predictions for the IM7/977 and T800/924 laminates would clearly have been non-conservative if the tails of their individual distributions represent real behaviour. Further investigation of this phenomenon is clearly called for.

One of the characteristics of the Weibull model is its reproductive property (Bury, 1975). A consequence of this property is that for a population of results that is well modelled by the Weibull distribution certain other features of the population, such as the minimum extreme values, will also be described by a Weibull distribution (Castillo, 1988). Or, more formally, the exact distribution of the smallest observations in sets of data that are described by a Weibull model also fits a Weibull model. This issue is of some significance to the designer, since he is more likely to be concerned about minimum-life values rather than median-life values which are most commonly used in modelling fatigue behaviour. Nakayasu (1987) has also shown that the Weibull shape factors for a number of related fatigue life data sets may themselves be modelled by a Weibull distribution. For a large number of data sets for a medium-carbon steel, for example, he obtained a good fit to a three-parameter Weibull model with the median value of the shape parameter being 1.69. The results of a similar such exercise are

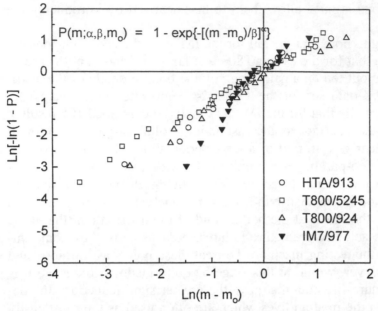

Figure 14. Three-parameter Weibull plot of the Weibull shape factors, m, for constant-stress fatigue tests on four CFRP laminates.

Table 1. Weibull parameters for the three-parameter Weibull distributions of the shape factors, m, for median fatigue life data for four CFRP laminates.

Material	Location parameter m_o	Shape parameter α	Correlation coefficient r
T800/5245	0.31	1.12	0.986
T800/924	0.20	0.96	0.987
IM7/977	0.10	1.76	0.970
HTA/913	0.33	1.12	0.986

shown in Figure 14, where the Weibull shape parameters, m, for the entire collection of results obtained for the four CFRP laminates are shown as three-parameter Weibull distributions.

The Weibull parameters for these data sets are given in Table 1 (only the shape and location parameters are given since the scale parameter for a normalised distribution is unity, as the figure shows). The symbol α is used here for the shape parameter in order to avoid confusion with the usual parameter m.

It can be seen that the degree of fit to a linear curve of slope near to unity is very good for the two T800 laminates and the HTA/913, but the IM7/977 composite appears to differ from the other three in having a higher slope and a lower value of the location parameter, m_o. But the differences in numerical values obtained by linear regression analysis notwithstanding, it can still be seen that the majority of the data points for the IM7/977 also fall within the general scatter band for the four groups of data. Note that the average value of α in the table is 1.28. This method of analysis provides further confirmation of the similar patterns of fatigue behaviour observed in these four laminates, despite individual differences in character, such as the fibre properties and degree of fibre/resin adhesion, that we have already noted.

3.4 *Constant-life analysis*

A complete set of fatigue data for any given material will necessarily include information on the effects of both tension and compression loads on the fatigue life, and the conventional way of reporting such data is in the form of a constant-life, or Goodman diagram such as that shown schematically in Figure 2. A typical set of $\sigma/\log N_f$ curves for high-performance CFRP laminate, one of those discussed in the last section, is shown in Figure 15 (Gathercole et al., 1994). The curves are plotted in terms of peak stress as a function of life. It is apparent, however, that as the compression

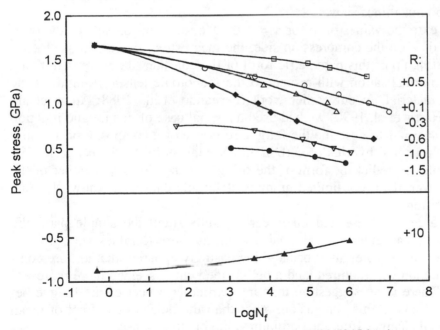

Figure 15. Median $\sigma/\log N_f$ curves for a $[(\pm45,0_2)_2]_s$ T800/5245 laminate at various R ratios.

component of cycling increases (R increasingly negative) the stress range, $2\sigma_{alt}$, to which the sample is subjected must at first increase and if the data were plotted as stress *range* versus loglife the data points for $R = -0.3$ would then apparently lie above those for $R = +0.1$. This is a familiar feature of fatigue in fibre composites, and it leads to the notion that some element of compression load in the cycle can apparently improve the fatigue response. It also results in a well-known aspect of composites fatigue, viz. that master diagrams of the constant life, or Goodman, variety are displaced from symmetry about the $R = -1$ plane (the alternating-stress axis) (Schütz & Gerharz, 1977; Kim, 1988). We have recently shown (Adam et al., 1992; Gathercole et al. 1994) that the effects of R ratio can be illustrated by presenting the fatigue data on a normalised constant life diagram by means of the function:

$$a = f\,(1 - m)^u\,(c + m)^v \tag{10}$$

where $a = \sigma_{alt}/\sigma_t$, $m = \sigma_m/\sigma_t$, and $c = \sigma_c/\sigma_t$. The alternating component of stress, σ_{alt}, is equal to $\frac{1}{2}\,(\sigma_{max} - \sigma_{min})$, and the mean stress, σ_m, is defined as $\frac{1}{2}\,(\sigma_{max} + \sigma_{min})$. σ_t and σ_c are the monotonic tensile and compressive strengths, respectively. For the purposes of this parametric analysis, we keep the sign of σ_c positive, so that the parameter c is also positive. The stress function, f, depends on the test material. Since this is a parabolic function, the criterion is more akin to the Gerber relationship than the normal Goodman linear law.

The extreme values of m for $a = 0$ are 1 on the tension side of the ordinate and $-c$ on the compression side: the mean stress range is thus $(1 + c)$. The setting up of this parametric form of the constant life curve is in fact a double normalisation with respect to the monotonic tensile strength of the hybrid material. In our earlier work (Fernando et al., 1988; Adam et al., 1989; Harris et al.,1990) we had used a special case of this fatigue function with $u = v = 1$ to describe the fatigue response of a composite of a family of unidirectional hybrids of carbon and Kevlar-49 in 914 epoxy resin. We were able to predict the form of the $\sigma/\log N_f$ curve for any member of the family even if only a limited amount of fatigue data were known for one composition.

But although these and other early results fitted the simple parabolic form for a particular life reasonably well, as more R values were investigated for more materials it became increasingly apparent that a more complex function was required, and a bell-shaped curve seemed more representative. There is no suggestion that any particular aspect of the fatigue behaviour leads to this conclusion, other than the deleterious effect of mean stress as it changes from the optimum value in either sense.

In order to apply this analysis, we use the following procedure:

1. After the replicate stress/life data are plotted, as in Figure 15, third-order polynomial curves are fitted, either to the median-life results or to the full data set, for each R ratio. A recent test of the difference between these two options results in negligible differences in the resulting life predictions, although it is possible that fitting the full data set is more useful in cases where the $\sigma/\log N_f$ curve is very flat.

2. The polynomial coefficients are extracted and inserted into a spread-sheet, together with established values of the monotonic tension and compression strengths, σ_t and σ_c. The spread sheet calculates the data pairs, (m, a), as defined by Equation 10, for each R value at predetermined values of life, N_f. The lives chosen for this usually fall within the experimental window, i.e. 10^4, 10^5 and 10^6 cycles.

3. The (m, a) data pairs are then plotted, together with the normalised values of tension and compression strength, which are, from the above definitions, unity and σ_c.

4. An analysis is carried out to establish the appropriate value of the parameter f. The simplest method is to use a non-linear regression method to fit the function in Equation 10 to the data sets for each life previously chosen, and observe the resulting values of f, u and v. Usually, the values of f are reasonably close to each other, and the goodness of fit does not change significantly if Equation 10 is refitted to the m,a curves with the value of f fixed at the mean value for the previous unconstrained fitting operations.

5. With the value of f fixed, the variation of u and v with life is then established in the form of a pair of functions:

$$u, v = A + B \log N_f \qquad (11)$$

These functions, with the empirically determined values of A and B are then used in a programme such as MathCAD to generate a family of predicted constant life curves such as those shown schematically in Figure 16. It is a simple matter to obtain curves for any life within the original experimental window and for any required R value. The proviso is that although some extrapolation to longer or shorter lives is acceptable if there is some indication from the $\sigma/\log N_f$ original data as to the direction that a particular stress/life curve may take, the extrapolation is only as good as the fit of the original polynomial curve to the data set.

A typical constant-life diagram, generated in the manner described above, is shown in Figure 17. The full curves have been obtained by allowing all three adjustable parameters, u, v and f to vary freely during the fitting process which, in this work is done with the aid of Microcal's Origin software. The coefficients of variation for most of these fitting operations are always of the order of 10% (standard error $\approx$ 2%).

Figure 17 also shows dashed curves which represent the results of sepa-

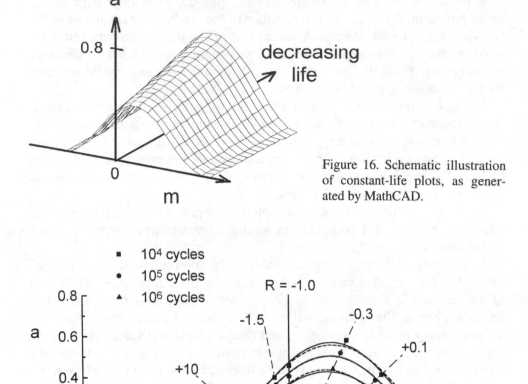

Figure 16. Schematic illustration of constant-life plots, as generated by MathCAD.

Figure 17. Constant-life plots for IM7/977 composite of $[(\pm45,0_2)_2]_s$ lay-up. The curves represent the relationship: $a = f\,(1 - m)^u\,(c + m)^v$. For the full curves, the three parameters are fitted without constraint, while for the dashed curves, the value of the parameter f has been fixed at the average value of 1.3 obtained by free-fitting.

rate fitting operations for which the value of f was fixed at the average of the three values obtained by fitting without constraint, in this case 1.3. It can be seen that this does not significantly impair the goodness of fit. We have applied this analysis to all of the composite laminates referred to in Figure 13 and the associated discussion. Of these, the T800/5245, T800/924 and IM7/977 composites are high-strength, intermediate-modulus fibres with similar characteristics, whereas HTA/913 is typical of the older-established lower strength and stiffness T300 composites. The basic me-

Table 2. Mechanical properties of four $[(\pm45,0_2)_2]_s$ composite laminates.

Material	Elastic modulus GPa	Tensile strength GPa	Failure strain %	Compression strength GPa
T800/5245	94 (3)	1.67 (0.9)	1.69 (0.1)	0.88 (0.1)
T800/924	92 (8)	1.42 (0.09)	1.49 (0.1)	0.90 (0.09)
IM7/977	90 (11)	1.43 (0.07)	1.52 (0.15)	0.90 (0.07)
HTA/913	70 (4)	1.27 (0.05)	1.73 (0.1)	0.97 (0.08)

Table 3. Values of the constant-life curve-fitting constants, f, u and v, for four $[(\pm45,0_2)_2]_s$ CFRP laminates.

Material	Life, log (cycles)	f	u	$sd\ (u)$	v	$sd\ (v)$
T800/5245	4	2.08	3.02	0.16	2.59	0.09
	5	2.08	3.37	0.13	2.77	0.06
	6	2.08	3.68	0.05	3.19	0.02
T800/924	4	1.30	2.01	0.11	2.15	0.11
	5	1.30	2.35	0.12	2.52	0.13
	6	1.30	2.83	0.15	2.94	0.16
IM7/977	4	1.30	1.97	0.05	2.23	0.05
	5	1.30	2.24	0.05	2.45	0.04
	6	1.30	2.64	0.05	2.73	0.04
HTA/913	4	1.00	2.13	0.11	2.50	0.13
	5	1.00	2.62	0.14	3.18	0.17
	6	1.00	3.34	0.18	4.32	0.23

chanical properties of the four laminates are shown in Table 2. Despite the fibre and resin differences, these materials all have very similar compression strengths, and the major effect of the resin and/or interface can be clearly seen in the differences between the two T800 composites.

Results of the constant life analysis are shown in Table 3 which gives values of f, u and v for all four laminates, and the dependence of u and v on life, as indicated in Equations 11, is shown in Figure 18.

Interpretation of the experimental values of f, u and v in Table 3 and Figure 18 is important in the context of evaluating the fatigue design potential of these laminates relative to one another. It follows from the derivation of the constant-life plots that the higher the values of f, the better the fatigue performance at any given life since f, overall, determines the relative 'height' of the curve, and the higher the curve, the greater the alternating stress that can be tolerated for a given mean stress at a given life. Likewise, the higher the values of u and v, the poorer the fatigue performance because the further u and v rise above unity (the parabolic 'special case' of the gen-

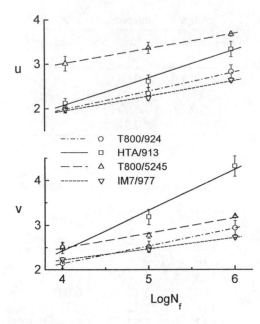

Figure 18. Dependence of the constant-life parameters u and v on life for four $[(\pm 45, 0_2)_2]_s$ CFRP laminates.

eralised constant-life relationship) the more the 'wings' of the curve are pulled downwards and the more bell-shaped the curve becomes, so *reducing* the level of alternating stress that can be tolerated for a given mean stress at a given life. It rather appears that the high v values for HTA/913 are associated with the high ratio of the monotonic compression and tensile strengths, σ_c/σ_t (i.e. the parameter c), since the HTA/913 laminate stands out from the others in this respect. And the greater the slopes of these lines, du/d (log N_f) and dv/d (log N_f) the poorer the fatigue performance because the higher the slope, the greater the downward deviation of the $\sigma/\log N_f$ curve at long lives. It will be appreciated that the greater the difference in the values of u and v for a particular material, the greater the degree of asymmetry of the constant-life curve. This may influence the choice of material if it is known that, for a given application, a relative degree of compression or tension loading will predominate.

With specific reference to these four laminates, it appears that there are relatively small differences in the actual values of u and v for all of the higher-performance laminates, although the T800/5245 constant-life plots are more asymmetric than those for the other two high-performance materials. But by and large there is little difference between the slopes, $du/d \log N_f$ and $dv/d \log N_f$, for these three materials. By contrast, the HTA/913 results stand out from the rest because the slopes are visibly much higher than those for the higher-performance laminates, a clearer indication of poorer fatigue performance than we obtain from a direct comparison of the raw stress/life data.

This model appears, from our investigations so far, to be applicable to a wide variety of types of CFRP and other kinds of composite. The apparently similar patterns of behaviour for different materials suggest that preliminary predictions of life for specific R ratios can be made on the basis of relatively small amounts of information, and that these preliminary predictions can be gradually improved as more fatigue experiments are carried out. For example, after we had obtained a full set of constant-life plots for the T800/924 laminate, we then began work on the IM7/977 material. We first obtained the monotonic tension and compression strengths and an σ/log N_f curve at $R = 0.1$ defined by about five median-life data points. We then produced a preliminary constant-life plot with these three fixed points and, by assuming that the f, u and v values for the material would be similar to those for T800/924, we then derived a set of predicted σ/log N_f curves at four other R ratios. When the actual experimental results were subsequently obtained for those same R ratios, it was found that although the true f, u and v values for the material were slightly different, the predicted curves fitted the preliminary estimates precisely in the case of the R ratios -1, -1.5 and $+10$, and sufficiently closely in the case of $R = 0.1$ and -0.3 to have allowed conservative design. The important question for any life prediction method of this kind is that which relates to the minimum amount of experimental data that is required in order for the designer to have confidence in any predicted life value.

3.5 *Damage accumulation and block-loading experiments*

There have been several studies of the accumulation of damage in a variety of composites, and arising out of these there have been proposed a number of models that permit fatigue life predictions to be made, with varying degrees of success. A recent survey (Barnard & Young, 1986) of life-prediction methods suggests that most existing models are limited by dubious or unreasonable assumptions or by lack of real knowledge of the actual degradation mechanisms (or at least of the effect of the accumulating damage on residual properties), especially at the low-stress/high-cycles end of the stress/log life (σ/log N_f) curve.

As a basic model for the prediction of life and residual properties for constant amplitude cycling, the strength/life equal-rank assumption used by many workers is reasonably successful and in principle holds out hope for its application to component testing since the predictions are made on the basis of relatively limited amounts of data, although most current methods derive from small-specimen laboratory-based techniques and must inevitably be followed up by component testing. But for variable amplitude cycling the situation in far less satisfactory. Attempts to predict behaviour following simple or complex block-loading patterns have usually given non-conservative results (Schütz & Gerharz, 1977) and have established that linear cumulative damage rules are inappropriate in most circumstances (Howe & Owen, 1972; Broutman & Sahu, 1969; Gerharz, 1982). In particular, Schütz and Gerharz observed that for certain types of laminate the linear damage rule predicted lives up to three times those actually measured – too large a margin for safe design work – and they suggest that this could be due to the very detrimental damage contributed by low-load cycles in the compression region which is not accounted for adequately in a Miner estimate.

We have recently carried out block-loading experiments on some high-performance composites in an attempt to derive a viable life prediction method (Adam et al., 1994) and like other workers we have met with difficulties that are so far unresolved. For example, we have tested samples of $[(\pm 45,0_2)_2]_s$ T800/5245 laminate, for which conventional constant-load fatigue data has been presented earlier, in four-unit block loading sequences. For the first four-unit sequence, an all-tension group, with $R = +0.1$, the four stress levels chosen were 1.3, 1.2, 1.1, and 1.0 GPa, indicating the narrow stress band between the monotonic strength and a notional 'fatigue limit'. Ideally, the block should have repeated itself several times before failure occurred, and the numbers of cycles at each of the stress levels were therefore chosen so that the block should account for 20% of the lifetime of the specimen if Miner's rule were to be obeyed. Thus, each unit in the sequence should have contributed 5% of the total damage to failure, and the block should have repeated about five times. The choice of 5% of the median life at each of the four stress levels was to avoid any possible overlap of lifetimes resulting from the scatter of results at each of the stress levels. Since the contribution to the total life of the specimen by each of the units within the block programme is 5% of the median life, each block of four units therefore had a notional Miner number, M, of 0.2, i.e.

$$\frac{p_1}{m(N_1)} + \frac{p_2}{m(N_2)} + \frac{p_3}{m(N_3)} + \frac{p_4}{m(N_4)} = M = 0.2 \qquad (12)$$

where p_i is the number of cycles in the unit, $m(N_i)$ is the median life at that particular stress level and M is the Miner number. From this it can be seen that if Miner's rule is followed, the block should repeat itself 5 times before failure occurs. Five block-loading tests were run for a given sequence of the stress units 1, 2, 3, 4, and of the 24 possible combinations six were selected for investigation.

Following the first all-tension (TTTT) series, the next stage was to introduce a compressive element into the loading sequence, the replacement of a unit in the block sequence being related to a factor common to both the old and the new units. The complete TTCT sequence involved testing six different combinations of these four units, as for the TTTT blocks.

The numbers of individual tests in these block-loading experiments are small by normal fatigue testing standards but one of the ways in which confidence in the results of small samples can be improved, however, is by pooling data, as discussed earlier. It is logical to assume that if the life data for a given constant stress level are randomly distributed then the Miner sum, M, defined by Equation 12, should be similarly distributed, regardless of whether the mean value or median value of M is equal to unity (Yang & Jones, 1981; Castillo et al., 1985). The most direct comparison that can be made between these two groups of test results (TTTT and TTCT), therefore, is to pool the Miner sums from all six groups for each sequence and present them as a Weibull plot, as in Figure 19. The populations are both well fitted by the three-parameter model with correlation coefficients greater than 0.98 (30° of freedom). The shape parameters, m, for the pooled M population, as shown on the graph, are 2.1 and 1.1, respectively, for the TTTT and TTCT groups. But while there is a marked difference in the level of variability between the all-tension and the mixed T/C tests, there is also a very real difference in the mean values of the Miner number for the two groups of tests,

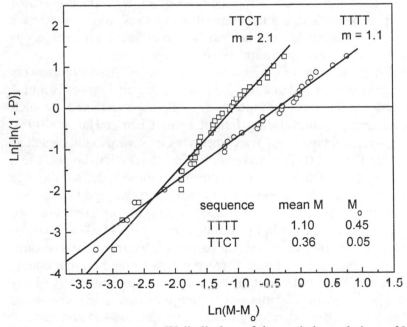

Figure 19. Three-parameter Weibull plots of the pooled populations of Miner numbers for four-unit block-loading tests on [(±45,0₂)₂]ₛ T800/5245 laminate. T = tension unit; C = compression.

1.1 for the TTTT sequence, and 0.36 for the TTCT sequence. To all intents and purposes, then. the Miner-Palmgren rule is valid for all-tension block-loading sequences for this material, and there is little doubt of the highly damaging consequence of introducing a single compression block into an otherwise all-tension group. Miner's law is no longer obeyed, by a substantial margin, the mean life having been reduced by some 60% as a result of the substitution. Similarly, although with less adequate statistical support, our results so far suggest that the substitution of a repeated tension unit into an all-compression sequence also resulted in a marked reduction in the mean Miner sum. Much more testing is required before the indications provided by these results can be accepted with confidence. But it appears reasonably sure, within the constraints imposed on interpretation of data subject to statistical variability, that whereas all-tension block sequences affect this particular composite in a fashion more or less describable by the linear damage rule, any sequence in which there is a change of stressing mode from tension to compression, or vice versa, results in immediate and drastic impairment of fatigue response. It appears to be the stress reversal as such which is damaging rather than the actual sequence or, with certain reservations, the actual stress levels.

Detailed analysis of the results of these block-loading experiments suggested the validity of a power law for fractional damage in terms of fractional life. From four-unit block loading the damage index in the power law is apparently stress dependent, being proportional to stress to the power 1.1 (i.e. approximately obeying Miners law, as we have seen) for tensile cycling and to the power -31.4 for compressive loading.

One of the difficulties that has arisen in the course of these experiments is that of reconciling the iso-damage model described in Figures 3 and 4 with the physical evidence of accumulation of damage obtained by such procedures as edge replication during fatigue tests. Chen & Harris (1993), for example, showed damage maps for composites of various kinds, such as those illustrated in Figure 20. The curves labelled 1-5 identify the point in a constant-load fatigue test when the first occurrence of the specified damage mechanism is observed. The supposition is that each damage-mechanism curve, including the final failure curve (σ/log N_f curve), represents part of an S-shaped decay curve of the kind postulated by Talreja (1981), although in the experimental window we see only a part of each curve. It is also supposed that at sufficiently low stresses, corresponding to the notional endurance limit, all curves will flatten out and converge at large numbers of cycles. It is clear, however, that for two given fatigue stress levels, the ratios of lives for which the various damage mechanisms start are not constant, so that the physical damage map cannot be reconciled with the concept of iso-damage lines described by the schematic illustration in Figure 3.

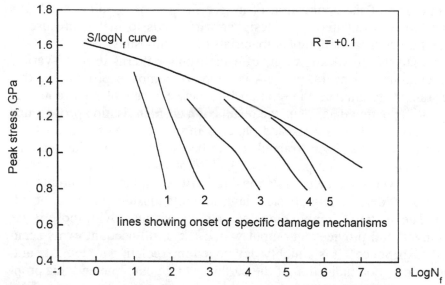

Figure 20. Schematic damage mechanism maps for T800/5245 [(±45,0₂)₂]ₛ laminate tested in repeated tension fatigue. The mechanisms identified in the graph are as follows: 1 = fibre fracture in 0° plies; 2 = matrix cracking in outer 45° plies; 3 = fibre drop out; 4 = matrix cracking in inner 45° plies; 5 = delamination at 0/45 interfaces.

3.6 *The application of neural network analysis*

Although a high level of understanding of the mechanisms of fatigue damage accumulation and the effects of this damage on the ability of a material or structure to sustain the service loads for which it was originally designed is of importance to materials developers, it must be said that the designer himself has little interest in the actual physical mechanisms of degradation. What he needs is simply a way of analysing available data to predict the likelihood of failure under a specified set of conditions.

In the last few years artificial neural networks have emerged as a new branch of computing suitable for applications in a wide range of fields. Artificial neural networks (ANNs) were developed originally to solve pattern recognition problems but their use has now been adopted in such fields as process control, failure analysis, non-destructive evaluation, and very many others. They appear to offer a means of dealing with many multi-variate properties for which an exact analytical model does not exist or at least is very difficult and time-consuming to develop. Fatigue seems to be just the type of materials property that is suitable for ANN analysis. Neural networks provide a compact way of coping with the large amounts of characterisation data generated in the study of a multi-variate dependent property such as fatigue. They also provide a very simple means of assessing the likely outcome of the application of a specified set of conditions. This is precisely what is required by a designer who needs to make safe use of complex fatigue data for complex materials like composites.

ANN analysis provides a means of establishing patterns in multi-variate experimental data. A neural network is a type of computer programme that can be 'taught' to emulate the relationships between sets of input data and corresponding output data. Thus the results of a characterisation programme can be 'learnt' by the network and patterns in the data can be established. Once the network has been 'trained' it can be used to predict the outcome of a particular set of input parameters. Unlike other types of analysis there is no need to know the actual relationship between inputs and outputs and the question of determining physical laws and relationships need not be addressed. The network simply compares patterns of inputs with those it has been 'taught' and provides an output which takes full account of its accumulated 'experience'. It is a highly developed interpolation system. Its outputs provide a good indication of the certainty of an outcome and this property may well prove to be a valuable indicator of where additional experimental information is required. The ability to signal the 'certainty' of the outcome is particularly important for materials like composites for which the fatigue response is often highly variable even when the monotonic strength properties of the same material may be quite closely definable.

Fatigue in composites involves many variables, including a variety of materials characteristics (e.g. fibre/resin mix, lay-up, moisture content) as well as the specific fatigue variables (mean and alternating stresses, variable stress characteristics, etc.), all of which may be treated as inputs to an ANN. Interestingly, one of the first questions that has arisen in our work in this field relates to the question of what constitutes failure. It is more likely to be the minimum life in a data set than the median or mean life, and it is important to decide whether the input should simply be the lowest value of life recorded in a data set, or some value estimated from the application of extreme value theory (Castillo, 1988). An essential issue is then to decide the exact nature of the distribution of fatigue lives in large and small samples.

4 ACKNOWLEDGEMENTS

Although much of this paper is in the nature of a review, it also makes much use of recent and current research at the University of Bath. It will be apparent from the references that this is the work of a considerable team of researchers, but I am particularly anxious to acknowledge the contributions and ideas of my current co-workers, Mr T. Adam, Dr D.P. Almond, Mr N. Gathercole, Mr J.A. Lee and Mr H. Reiter. We are grateful to the Defence Research Agency and the Engineering and Physical Sciences Research Council of the UK for financial support of much of the research referred to in the paper.

REFERENCES

Adam, T., Dickson, R.F., Jones, C.J., Reiter, H. & Harris, B. (1986). *Proc. Inst. Mech. Engrs.: Mech. Eng. Sci.*, 200, C3, 155-166.

Adam, T., Fernando, G., Dickson, R.F., Reiter, H. & Harris, B. (1989). *Int. J. Fatigue,* 11, 233-237.

Adam, T., Gathercole, N., Reiter, H. & Harris, B. (1992). *Advanced Composites Letters,* 1, 23-26.

Adam, T., Gathercole, N., Reiter, H. & Harris, B. (1994). *Int. J. Fatigue,* 6, 533-548.

Barnard, P.M. & Young, J.B. (1986). Cumulative fatigue and life prediction of composites, Final Report on MoD contract 11494/2080-0134 XR/MAT, Cranfield Institute of Technology, November 1986.

Beaumont, P.W.R. & Harris, B. (1972). Carbon Fibres: Their Composites and Applications. (Plastics Institute (now Plastics and Rubber Institute). London), 283-291.

Broutman, L.J. & Sahu, S. (1969). *Proceedings of the 24th Annual Technical Conference of the Reinforced Plastics/Composites Institute of SPI*, paper 11D, (SPI, New York).

Bunsell, A.R. (1975). *J. Mater. Sci.* 10, 1300-1308.

Bury, K.V. (1975). Statistical Models in Applied Science, (J. Wiley & Sons, London), Chapters 11 and 12.

Castillo, E., Alvarez, E., Cobo, A. & Herrero, T. (1993). An Expert System for the Analysis of Extreme Value Problems, University of Cantabria, based on the book Extreme Value Theory In: Engineering, E. Castillo, (1988), (Academic Press, Boston, London).

Castillo, E., Canteli, A.F., Esslinger, V. & Thürlimann, B. (1985). Statistical Model for Fatigue of Wires, Strands and Cables, *IABSE Periodical, International Association for Bridge and Structural Engineering* (Zurich, Switzerland).

Chatfield, C. (1970). Statistics for Technology. Penguin, Harmondsworth, Middx.

Chen, A.S. & Harris, B. (1993). *J. Mater. Sci,* 28, 2013-2027.

Dobb, M.G., Johnson, D.J., Majeed, A. & Saville, B.P. (1979). *Polymer,* 20, 1284-88.

Fernando, G., Dickson, R.F., Adam, T., Reiter, H. & Harris, B. (1988). *J. Mater. Sci.,* 23, 3732-3743.

Fernando, G., Dickson, R.F., Adam, T., Reiter, H. & Harris, B. (1988). *J. Mater. Sci.,* 23, 3732-3743.

Fuwa, M., Harris, B. & Bunsell, A.R. (1975). *J. Phys. D.* (Applied Physics), 8, 1460-1471.

Gathercole, N., Reiter, H., Adam, T. & Harris, B. (1994). *Int. J. Fatigue,* 16, 523-532.

Gerber, W.Z. (1874). *Z. Bayer Archit. Ing. Ver.,* 6, 101.

Gerharz, J.J. (1982). Practical Considerations of Design, Fabrication and Tests for

Composite Materials, *AGARD Lecture Series No. 124*, Series Director B. Harris, paper 8. AGARD/NATO, Paris.

Goodman, J. (1899). *Mechanics Applied to Engineering.* Longman Green: London

Harris, B. (1994). Fatigue of glass-fibre composites, Handbook of Polymer-Fibre Composites, ed. by F.R. Jones. Longman, Harlow, UK, 309-316.

Harris, B., Reiter, H., Adam, T., Dickson, R.F. & Fernando, G. (1990). *Composites*, 21, 232-242.

Harris, B., Reiter, H., Adam, T., Dickson, R.F. & Fernando, G. (1990). *Composites*, 21, 232-242.

Howe, R.J. & Owen, M.J. (1972). *Proceedings of the Eighth International Reinforced Plastics Congress (BPF, London),* 137-148.

Jones, C.J., Dickson, R.F., Adam, T., Reiter, H. & Harris, B. (1984). *Proc. Roy. Soc. Lond.,* A396, 315-338.

Kim, R.Y. (1988). Composites Design (4th Ed.), S.W. Tsai, Editor. Think Composites, Dayton, Ohio, USA, Chapter 19.

Lamela-Rey, M.J. (1994). Proceso de Acumulacion de Daño en Flexion para Laminados Simetricos de Fibra de Carbono, Doctoral thesis of the Departamento de Construccion e Ingenieria de Fabricacion, University of Oviedo at Gijon, Spain.

Mandell, J.F. (1982). Developments in Reinforced Plastics-2, G. Pritchard (ed.), 62-108. Applied Science Publishers, London.

Marco, S.M. & Starkey, W.L. (1954). *ASME Trans.*, 76, 627.

Miner, M.A. (1945). *J. Appl. Mech.*, 12, A159.

Nakayasu, H. (1987). Statistical analysis of small-sample fatigue data. Statistical Research on Fatigue and Fracture, *Current Japanese Materials Research vol. 2*, The Society of Materials Science, Japan, Eds T. Tanaka, S. Nishijima & M. Ichikawa. Elsevier Applied Science, London, 21-43.

Owen, M.J. & Morris, S. (1971). Carbon Fibres: Their Composites and Applications. Plastics Inst., London, 292-302.

Reifsnider, K.L. & Jamison, R.D. (1982). *Int. J. Fatigue*, 4, 187-198.

Ryder, J.T. & Walker, E.K. (1977). Fatigue of Filamentary Composite Materials, *ASTM STP 636*. American Society for Testing and Materials, Philadelphia, USA, 3-26.

Schütz, D. & Gerharz, J.J. (1977). *Composites*, 8, 245-250.

Sims, G.D. & Gladman, D.G. (1978). *Plastics and Rubber: Materials & Applications*, 1, 41-48.

Sturgeon, J.B. (1975). RAE (Farnborough) Technical Report MAT 75315. UK Ministry of Defence Procurement Executive.

Talreja, R. (1981). *Proc. R. Soc. (Lond),* A378, 461-475.

Weibull, W. (1959). Fatigue Testing and Analysis of Results. Pergamon Press, NY.

Whitney, J.M. (1981). Fatigue of Fibrous Composite Materials. *ASTM STP 723.* American Society for Testing and Materials, Philadelphia, USA, 133-151.

Yang, J.N. & Jones, D.L. (1981). Fatigue of Fibrous Composite Materials, *ASTM STP 723.* American Society for Testing and Materials, Philadelphia, USA, 213-232.

Time-dependent phenomena related to the durability analysis of composite structures

I. Emri
University of Ljubljana, Ljubljana, Slovenia

ABSTRACT: The mechanical and other physical properties of polymer-based composites can significantly change with time and this can seriously influence their long term durability. This time-dependent change in the materials' mechanical properties is caused either by a chemical process, or arises from the viscoelastic nature of the polymer matrix. The paper reviews the experimental techniques and the mathematical formalism needed for the characterization and modelling of the time-dependent mechanical material functions, discusses limitations of the existing theories (models), and presents some latest developments in the mathematical modelling of viscoelastic material functions.

It then discusses physical ageing, the changes in the mechanical behaviour that may occur as a consequence of temperature and pressure variations in the glassy state. Since the behaviour of composite materials is generally non-linear, the discussion ends with a review of some existing non-linear models.

1 INTRODUCTION

Polymer based composites are becoming increasingly important structural materials. Their mechanical, as well as other physical properties, can significantly change with time and this can seriously influence their long term durability. These time-dependent changes in the materials' mechanical properties is caused either by a chemical process, or is due simply to the viscoelastic nature of the polymer matrix. In this paper the emphasis will be on the latter phenomenon.

The experimental methods and the mathematical formalism needed for the characterization and modelling of the time dependent mechanical material functions will be reviewed in Section 2.

Certain models and their limitations will be discussed in Section 3. Section 4 will deal with generalized stress-strain relations. Section 5 discusses the role of relaxation and retardation spectra in the characterization of viscoelastic behaviour.

Some latest developments in the mathematical modelling of viscoelastic material functions by discrete spectra will be presented in Section 6.

The separate and combined influence of temperature, pressure, and moisture on the long term mechanical properties will be analyzed in Section 7. It will be shown that variations in pressure and moisture history can yield similar time-dependent response as variations in temperature history. The influence of temperature and pressure variations to which the material is exposed during the manufacturing process or during later applications can cause time-delayed changes in the mechanical and other physical properties. This phenomenon, known as physical ageing, can substantially influence the long term durability of a material. Understanding of these phenomena is of great practical importance. It will be discussed in Section 8. Section 9 deals with multiaxial stress states. The behaviour of composite materials under any but the smallest strains is generally non-linear. Some existing non-linear viscoelastic models (theories) will be reviewed briefly in Section 10. The paper concludes with an introduction to viscoelastic stress analysis in Section 11.

2 MATERIAL CHARACTERIZATION

Experimentally, one seeks to characterize materials by performing simple laboratory tests from which information relevant to actual in-use conditions may be obtained. In the case of viscoelastic materials, mechanical characterization often consists of performing uniaxial tensile tests similar to those used for elastic solids, but modified so as to enable observation of the time dependence of the material response. Although many such *viscoelastic*

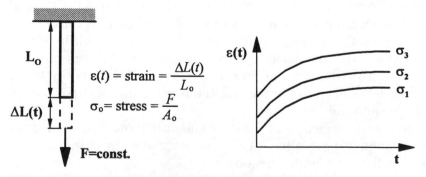

$$\varepsilon(t) = \text{strain} = \frac{\Delta L(t)}{L_0}$$

$$\sigma_0 = \text{stress} = \frac{F}{A_0}$$

Figure 1. Schematic of the uniaxial creep experiment.

tensile tests have been used, one most commonly encounters mainly creep, stress relaxation, and dynamic (sinusoidal) loading (stress or strain controlled).

2.1 Creep

The tensile creep test consists of measuring the time-dependent uniaxial strain, $\varepsilon(t)$, resulting from the application of a constant uniaxial stress (dead load), σ_o. Their ratio is the *creep compliance*,

$$D(t) = \varepsilon(t) / \sigma_o \tag{1}$$

If the response of the material is *linear,* the creep curves resulting from various applied stresses supply the same information. This is shown in Figure 2 where $D(t)$ as a function of time is presented schematically in both linear and logarithmic coordinates as a function of the time, t. D_g is the glassy (instantaneous) compliance, while D_e is the equilibrium (retarded) compliance.

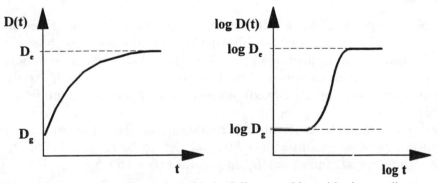

Figure 2. Creep compliance plotted in both linear and logarithmic coordinates.

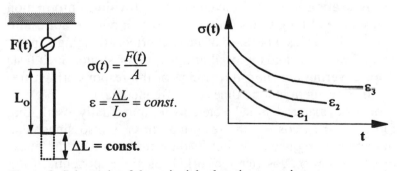

Figure 3. Schematic of the uniaxial relaxation experiment.

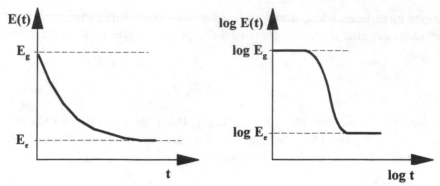

Figure 4. Relaxation modulus plotted in both linear and logarithmic coordinates.

2.2 *Stress relaxation*

Another common test consists of monitoring the time-dependent uniaxial stress, $\sigma(t)$, resulting from the imposition of a constant uniaxial strain, ε_o.

The ratio of the stress to the strain is the *relaxation modulus*, defined as

$$E(t) = \sigma(t) / \varepsilon_o \tag{2}$$

Again, if the response of the material is linear, the relaxation curves resulting from various applied strains furnish the same information as shown in Figure 4 where $E(t)$ as a function of time is presented schematically in both linear and logarithmic coordinates as function of the time, t. E_g is the glassy (instantaneous or unrelaxed) modulus and E_e is the equilibrium (relaxed) modulus.

Creep and relaxation are both manifestations of the same molecular mechanisms, and one should expect $E(t)$ and $D(t)$ to be related. However, even though $E_g = 1/D_g$ and $E_e = 1/D_e$, in general $E(t) \neq 1/D(t)$.

2.3 *Dynamic loading*

Creep and stress relaxation tests are convenient for providing information on the material response at long times (minutes to days), but are not useful at shorter times (seconds or less) because of inertial effects (ringing). In *dynamic tests* one applies a sinusoidally oscillating stress or strain. Such tests are well-suited for covering the short-time range of the response (high frequencies) but are inconvenient at long times (low frequencies).

When a viscoelastic material is subjected to a sinusoidally oscillating strain of the form $\varepsilon(t) = \varepsilon_o \cos \omega t$, the resulting stress is also sinusoidal, having the same angular frequency ω, but leading the strain by the phase angle $\delta(\omega)$. Hence $\delta(\omega) = \sigma_o \cos [\omega t + \delta(\omega)]$. Using the exponential form one may write

$$\varepsilon(\omega) = \varepsilon_o \exp i\omega t$$
$$\sigma(\omega) = \sigma_o(\omega) \exp i[\omega t + \delta(\omega)] \tag{3}$$

where $\sigma(\omega)$ and $\varepsilon(\omega)$ are the steady-state stress and strain, respectively, and $i = \sqrt{-1}$. The stress amplitude is a function of the frequency. The *complex modulus*, $E^*(\omega)$, is then defined as

$$E^*(\omega) = \sigma(\omega) / \varepsilon(\omega) = E'(\omega) + iE''(\omega) \tag{4}$$

$E'(\omega)$, the real part of $E^*(\omega)$, termed the *storage modulus*, is a measure of the material's ability to store energy elastically. The imaginary part, the *loss modulus*, $E''(\omega)$, is a measure of its ability to dissipate energy through viscous mechanisms, and is a parameter often related to the toughness and impact resistance of the material. In composite materials Coulomb friction between the matrix and the fibers can substantially increase $E''(\omega)$. A useful interpretation of the complex modulus is obtained by separating the stress into two parts, one in phase with the strain and the other $\pi/2$ out-of-phase. Then

$$\sigma'(\omega) = \sigma_o(\omega) \cos \delta(\omega) \exp i\omega t = \varepsilon_o E'(\omega) \exp i\omega t$$
$$\sigma''(\omega) = \sigma_o(\omega) \sin \delta(\omega) i \exp i\omega t = \varepsilon_o E''(\omega) i \exp i\omega t \tag{5}$$

The storage modulus is the ratio of the in-phase stress component, while the loss modulus is the ratio of the out-of-phase stress component, to the strain. Thus

$$E'(\omega) = \sigma_o'(\omega) / \varepsilon_o$$
$$E''(\omega) = \sigma_o''(\omega) / \varepsilon_o \tag{6}$$

where $\sigma_o' = \sigma_o(\omega) \cos \delta(\omega)$ and $\sigma_o'' = \sigma_o(\omega) \sin \delta(\omega)$. Relations between the various components of the complex modulus follow easily from the vector visualization (see Figure 5).

$$\tan \delta(\omega) = E''(\omega) / E'(\omega)$$
$$E^*(\omega) = \sqrt{[E'(\omega)]^2 + [E''(\omega)]^2}$$
$$E'(\omega) = E^*(\omega) \cos \delta(\omega) \tag{7}$$
$$E''(\omega) = E^*(\omega) \sin \delta(\omega)$$

The total work done per cycle by both the in-phase and out-of-phase components is:

$$W = \int_0^{2\pi/\omega} \sigma(t) \, d\varepsilon(t) = \int_0^{2\pi/\omega} \sigma(t) \, \dot{\varepsilon}(t) \, dt$$

and therefore:

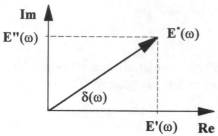

Figure 5. Components of the complex modulus in the complex plane.

$$W = \int_0^{2\pi/\omega} [\sigma'(\omega) \cos \omega t] (-\varepsilon_o \omega \sin \omega t) \, dt$$

$$- \int_0^{2\pi/\omega} [\sigma''(\omega) \sin \omega t] (-\varepsilon_o \omega \sin \omega t) \, dt \qquad (8)$$

$$= \pi \, \sigma_o'' (\omega)\varepsilon_o = \pi\sigma_o (\omega)\varepsilon_o \sin \delta(\omega)$$

Note that the in-phase components produce no net work when integrated over a cycle (the stored energy is 100% 'recoverable'), while the out-of-phase components result in a net dissipation per cycle equal to:

$$W_{\text{dis}} = \pi\sigma_o'' (\omega) \, \varepsilon_o = \pi\sigma_o(\omega)\varepsilon_o \sin \delta(\omega) \qquad (9)$$

Within a single cycle the maximum energy stored by the in-phase components occurs over a quarter cycle. This energy equals:

$$W_{\text{st}} = \int_o^{\pi/2\omega} [\sigma'(\omega) \cos \omega t] (-\varepsilon_o \omega \sin \omega t) \, dt$$

$$= \sigma_o' \, \varepsilon_o / 2 = \sigma_o(\varepsilon_o / 2) \cos \delta(\omega) \qquad (10)$$

The relative dissipation, the ratio $W_{\text{dis}}/W_{\text{st}}$, is then related to the phase angle by

$$\frac{W_{\text{dis}}}{W_{\text{st}}} = 2 \pi \tan \delta(\omega) \qquad (11)$$

and is known as the *specific loss*.

3 SPRING-AND-DASHPOT MODELS

The linear differential equations used to describe relaxation and retardation phenomena can be related to the behaviour of models composed of linear spring and dashpot elements. Such models are graphical representations of the differential equations and can greatly aid in visualizing the material response. In addition, they furnish equations that are convenient in automatic computation. The models are sometimes modified to represent even non-linear behaviour (see Section 10.2).

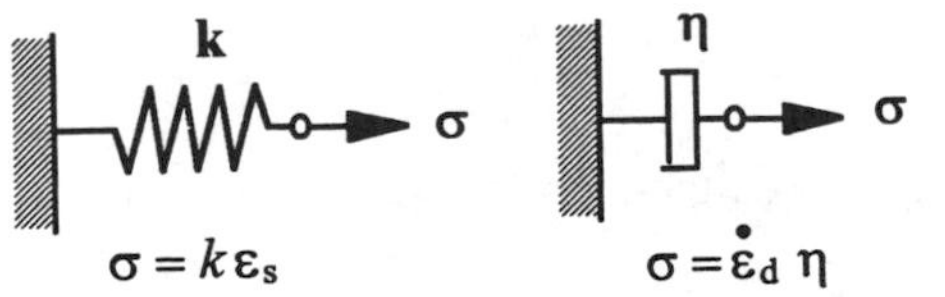

$$\sigma = k\varepsilon_s \qquad\qquad \sigma = \dot{\varepsilon}_d\,\eta$$

Figure 6. Hookean spring and Newtonian dashpot.

The spring-and-dashpot models are constructed from combinations of Hookean springs and Newtonian dashpots, defined as in Figure 6 above where k is the spring stiffness, η is the Newtonian viscosity, the dot indicates differentiation with respect to time, ε_s is the strain across the spring, and $\dot{\varepsilon}_d$ is the rate of strain across the dashpot.

3.1 *The Maxwell* unit

A series combination of a spring and a dashpot is called a *Maxwell unit*. Since the same stress acts through both elements of the unit and the strains are additive, we have

$$\dot{\varepsilon} = \frac{\dot{\sigma}}{k} + \frac{\sigma}{\eta} \tag{12}$$

This expression, which interrelates the stress, the strain, and the rate of strain, is the *constitutive equation* for the Maxwell unit. Note that it contains time derivatives, so that a simple constant of proportionality between stress and strain does not exist. Similarly, the concept of *modulus* – the ratio of stress to strain – must be broadened to account for this more complicated behaviour. One may define several different moduli appropriate for various types of loading (excitations). The relaxation modulus and the dynamic moduli discussed earlier were just two examples. It is usually not difficult to obtain these various moduli from the governing constitutive equation by solving it as an ordinary differential equation subject to the appropriate boundary conditions. In a stress relaxation test, for instance, one has $\varepsilon(t) = \varepsilon_o = \text{const}$, and therefore

$$\frac{1}{k}\frac{d\sigma}{dt} = -\frac{1}{\eta}\sigma$$

$$\int_{\sigma_o}^{\sigma} \frac{d\sigma}{\sigma} = -\frac{k}{\eta}\int_o^t dt \tag{13}$$

$$\sigma = \sigma_o \exp(-t/\tau)$$

in Equation (13), $\tau = \eta/k$ is a characteristic parameter with units of time, termed the *relaxation time*. The relaxation modulus, $E(t)$, may be obtained

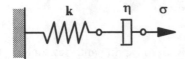

Figure 7. The Maxwell unit.

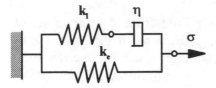

Figure 8. Model of the standard linear solid.

from Equation (13) directly, noting that the initial stress is just that needed to stretch the spring to a strain, ε_o. Thus,

$$E(t) = \frac{\sigma(t)}{\varepsilon_o} = \frac{\sigma_o}{\varepsilon_o} \exp(-t/\tau)$$

i.e.

$$E(t) = k \exp(-t/\tau) \tag{14}$$

No polymer actually exhibits the unrestricted flow permitted by the Maxwell unit which is called a *unit* because it is not a true viscoelastic model. It can represent stress relaxation but not strain retardation. A viscoelastic model must be able to represent both phenomena, albeit with a change of parameters. Placing a spring in parallel with the Maxwell unit remedies this deficiency and furnishes what is known as the model of the *standard linear solid* (SLS).

Now the same stress acts through the spring and the spring-and-dashpot combination so that $\sigma = \sigma_e + \sigma_1$ and the strains become $\varepsilon = \varepsilon_e = \varepsilon_1$ and $\varepsilon_1 = \varepsilon_{s1} + \varepsilon_{d1}$. Hence, the constitutive equation for this model is obtained as

$$(k_e + k_1)\dot{\varepsilon} + \frac{k_e \varepsilon}{\tau_1} = \dot{\sigma} + \frac{\sigma}{\tau_1} \tag{15}$$

and the relaxation modulus becomes

$$E(t) = k_e + k_1 \exp(-t/\tau_1) \tag{16}$$

as shown in Figure 9.

In the SLS model k_e and k_1 are chosen to fit the glassy and rubbery moduli, and τ_1 is chosen so as to place the relaxation in the correct time interval. The three-parameter SLS model is able to describe correctly the es-

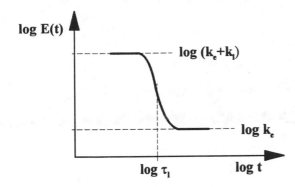

Figure 9. The relaxation modulus of the standard linear solid (SLS).

sential features of viscoelastic relaxation. However, the ability of the model to fit experimental data is poor. The transition it predicts is generally too abrupt.

3.2 The Wiechert model

In those cases where an improved fit to experimental data is necessary, one may introduce additional Maxwell units in parallel. Such a model is called a Wiechert model. As the number of elements increases, the determination of the stiffnesses and viscosities becomes more difficult. Section 6 describes a method that determines a Wiechert model from experimental data.

The model introduced above is most easily treated using operational calculus. Letting $D = \partial/\partial t$ be the time-derivative operator, one may write the stress transmitted by the jth Maxwell unit of the model as,

$$D\sigma_j + \frac{1}{\tau_j}\sigma_j = k_j D\varepsilon$$

and therefore

$$\sigma_j = \frac{k_j D\varepsilon}{(D+1/\tau_j)} \tag{17}$$

The total stress transmitted by the model is the sum of all the σ_j, plus the stress in the equilibrium spring k_e,

$$\sigma = k_e\varepsilon + \sum_j \sigma_j = \left(k_e + \sum_j \frac{k_j D}{D+1/\tau_j}\right)\varepsilon \tag{18}$$

Equation (18) represents the constitutive equation for the Wiechert model, although it is somewhat inconvenient to use in this form. For any strain as a

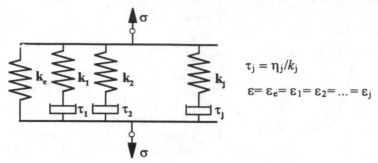

$$\tau_j = \eta_j/k_j$$

$$\varepsilon = \varepsilon_e = \varepsilon_1 = \varepsilon_2 = \dots = \varepsilon_j$$

Figure 10. The Wiechert model.

function of time, the stress output must be obtained by first clearing the operational equation of fractions and then solving the time-dependent differential equation by separation of variables or some other technique.

Instead of going through this tedious procedure one may utilize the convenient Laplace transform method to handle the operator equation (Tschoegl 1989). Using an overhead bar to denote the transform of a quantity and using the symbols to denote the transform variable, Equation (14) can be written in the transform plane as

$$\overline{\sigma}(s) = \left[k_e + \sum_j \frac{k_j s}{s + 1/\tau_j} \right] \overline{\varepsilon}(s) \tag{19}$$

The quantity in brackets is the Laplace transform of the modulus of the Wiechert model that relates the stress transform to the strain transform. We write

$$\Re(s) = \frac{\overline{\sigma}(s)}{\overline{\varepsilon}(s)} = k_e + \sum_j \frac{k_j s}{s + 1/\tau_j} \tag{20}$$

where $\Re(s)$, the ratio of the general stress to the general strain transform, is called the *response transform*. For a given strain excitation one obtains the resulting stress in three steps:

1. Obtain an expression for the transform of the strain function, $\overline{\varepsilon}(s)$;
2. Form the algebraic product $\overline{\sigma}(s) = \Re(s)\overline{\varepsilon}(s)$; and
3. Obtain the inverse transform of the result to yield the stress function in the time plane.

The advantage of the transform method is that the transforms and their inverses are tabulated extensively (see, e.g. Tschoegl 1989). Following this scheme for the stress relaxation test, $\varepsilon(t) = \varepsilon_o$ gives

$$\overline{\varepsilon}(s) = \varepsilon_o / s$$

$$\overline{\sigma}(s) = \Re(s)\,\overline{\varepsilon}(s) = \left[k_e + \sum_j \frac{k_j s}{s+1/\tau_j}\right]\frac{\varepsilon_o}{s}$$

and

$$\sigma(t) = L^{-1}\{\overline{\sigma}(s)\} = k_e + \left[\sum_j k_j \exp(-t/\tau_j)\right]\varepsilon_o \qquad (21)$$

where L^{-1} denotes the inverse Laplace transformation. The quantity in brackets is clearly just the time-dependent relaxation modulus for the Wiechert model.

It is useful to note that the relaxation modulus may be obtained directly from the response transform in accordance with

$$\overline{E}(s) = \frac{\overline{\sigma}}{\varepsilon_o} = \frac{1}{\varepsilon_o}\Re(s)\frac{\varepsilon_o}{s} = \frac{1}{s}\Re(s) \qquad (22)$$

This result is valid for a Wiechert model of any size.

As another example, consider the stress resulting from a constant-strain-rate test. Let R be a constant rate of straining. Then $\varepsilon(t) = Rt$ and we have

$$\overline{\varepsilon}(s) = R/s^2$$

$$\overline{\sigma}(s) = \Re(s)\overline{\varepsilon}(s) = \left[k_e + \sum_j \frac{k_j s}{s+1/\tau_j}\right]\frac{R}{s^2} =$$

$$= \frac{k_e R}{s^2} + \sum_j \frac{k_j R}{s(s+1/\tau_j)}$$

and retransformation gives

$$\sigma(t) = k_e Rt + R\sum_j k_j \tau_j[1 - \exp(-t/\tau_j)] \qquad (23)$$

The slope of the constant-strain-rate stress-strain curve is related to the relaxation modulus evaluated at time t by

$$\frac{d\sigma}{d\varepsilon} = \frac{d\sigma}{dt}\frac{dt}{d\varepsilon} = \frac{d\sigma}{dt}\frac{1}{R} = k_e + \sum_j k_j \exp(-t/\tau_j) = |E(t)|_{t=\varepsilon/R} \qquad (24)$$

In the case of the *dynamic modulus*, it is possible to write the result directly without going through the Laplace transformation. Since the time dependence of both the stress and the strain are of the form $\exp(i\omega t)$, all time derivatives will contain the expression $(i\omega)\exp(i\omega t)$. One may simply replace the **D**-operator in Equation (18) or the *s*-operator in Equation (19) by $i\omega$ and write

$$\sigma(\omega) = \left[k_e + \sum_j \frac{k_j i\omega}{i\omega + 1/\tau_j} \right] \varepsilon(\omega)$$

For the modulus we then obtain

$$E*(\omega) = \frac{\sigma(\omega)}{\varepsilon(\omega)} = k_e + \sum_j \frac{k_j(i\omega)}{i\omega + 1/\tau_j} \tag{25}$$

This equation can be manipulated algebraically to yield several useful alternative forms

$$E*(\omega) = \frac{\sigma(\omega)}{\varepsilon(\omega)} = k_e + \sum_j k_j - \sum_j \frac{k_j}{1 + i\omega\tau_j}$$

or, separating the real and imaginary parts,

$$E*(\omega) = k_e + \sum_j \frac{k_j \omega^2 \tau_j^2}{1 + \omega^2 \tau_j^2} + i \sum_j \frac{k_j \omega \tau_j}{1 + \omega^2 \tau_j^2} \tag{26}$$

One other application of the Wiechert model is worth discussing briefly: those instances in which the input strain function is not know as a mathematical expression, or where its mathematical expression is so complicated as to make the transform method intractable. In those cases, one may return to the differential constitutive equation and recast it in finite-difference form so as to obtain a numerical solution. Recall that the stress in the jth Maxwell unit of the Wiechert model is given by

$$\frac{d\sigma_j}{dt} + \frac{1}{\tau_j} \sigma_j = k_j \frac{d\varepsilon}{dt} \tag{27}$$

This can be written in finite difference form as

$$\frac{\sigma_j^t - \sigma_j^{t-1}}{\Delta t} + \frac{1}{\tau_j} \sigma_j^t = k_j \frac{\varepsilon_j^t - \varepsilon_j^{t-1}}{\Delta t} \tag{28}$$

where the superscripts *t* and *t*–1 indicate values before and after the passage of a small time increment Δt. Solving for σ_j^t yields:

$$\sigma_j^t = \frac{1}{1 + (\Delta t / \tau_j)} [k_j(\varepsilon^t - \varepsilon^{t-1}) + \sigma_j^{t-1}] \tag{29}$$

Summing over all Maxwell units and adding the stress in the equilibrium spring gives

$$\sigma^t = k_e \varepsilon^t + \sum_j \frac{k_j (\varepsilon_j^t - \varepsilon_j^{t-1}) + \sigma_j^{t-1}}{1 + (\Delta t / \tau_j)} \tag{30}$$

This constitutes a recursive algorithm that can be used to calculate successive values of σ^t beginning at $t = 0$.

3.3 *The Voigt unit and Kelvin model*

The Maxwell and Wiechert models are useful for relaxation-type loadings where the stress output resulting from a strain input is required, but they are awkward for creep loadings where the stress is the input function. For modelling the creep response the appropriate models are the Voigt unit and the Kelvin model. The first consist of a spring and a dashpot in parallel. Its characteristic time, τ, is the *retardation time*. The second combines several Voigt models in series with an isolated spring and/or dashpot. The reader can easily derive the corresponding constitutive relations, following the steps used above in the derivation of the constitutive relations for the Maxwell and the Wiechert models (for details see Tschoegl 1989).

3.4 *Fitting models to experimental data*

The material constants (the k's and τ's) that appear in the model formulations must be chosen by reference to appropriate experimental data. Several methods are available for accomplishing this. There are some older ones (see e.g. Tschoegl 1989) that may be useful at times. More recently, however, the easy availability of considerable computer power has sprouted a number of methods that determine a distribution of relaxation (or retardation) time distributions (Baumgaertel & Winter 1989, 1992; Honercamp 1989; Honercamp & Weese 1989, 1993; Elster et al. 1991). The algorithm developed by Emri & Tschoegl (1992, 1993, 1994) is a recursive computer algorithm that obtains a set of moduli corresponding with a set of preselected equally spaced relaxation times (or compliances and retardation times).

Due to the complexities of working with a distribution of relaxation or retardation times various investigators have proposed simple empirical mathematical models to fit experimental data. This topic has been discussed extensively by Tschoegl (1989).

4 GENERALIZED STRESS-STRAIN RELATIONS

Although the spring-and-dashpot models have their uses, this approach is somewhat restrictive. More general treatments, although not always useful in engineering practice, are commonly encountered in the literature. For completeness, some extensions of the spring-and-dashpot models to these more general formulations will be described briefly.

4.1 *Differential equation formulation*

Recall the operational form of the constitutive relation for the Wiechert model (Equation (18)),

$$\sigma = \left(k_e + \sum_j \frac{k_j D}{D + 1/\tau_j} \right) \varepsilon$$

If the expression in brackets is expanded and recast with a common denominator, it takes the form:

$$\sigma = \frac{b_m \dfrac{\partial^m}{\partial t^m} + \dots + b_1 \dfrac{\partial}{\partial t} + b_0}{a_n \dfrac{\partial^n}{\partial t^n} + \dots + a_1 \dfrac{\partial}{\partial t} + a_0} \tag{31}$$

This higher order differential equation relating the stress to the strain is sometimes taken as a starting point for developing the theory of linear viscoelasticity.

For the two-unit Wiechert model, for instance,

$$\sigma = \left(k_e + \frac{k_1 D}{D + 1/\tau_1} + \frac{k_2 D}{D + 1/\tau_2} \right) \varepsilon \tag{32}$$

This equation has the common denominator

$$(D + 1/\tau_1)(D + 1/\tau_2) = D^2 + (1/\tau_1 + 1/\tau_2)D + 1/(\tau_1 \tau_2)$$

so the operator equation may be written as

$$\sigma = \frac{k_g D^2 + [(k_e + k_1)/\tau_1 + (k_e + k_2)/\tau_2]D + k_e/\tau_1 \tau_2}{D^2 + (1/\tau_1 + 1/\tau_2)D + 1/(\tau_1 \tau_2)} \tag{33}$$

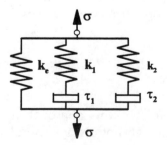

Figure 11. The two-unit Wiechert model.

where

$$k_g = k_e + k_1 + k_2$$

This equation is of the form

$$a_2 \frac{\partial^2 \sigma}{\partial t^2} + a_1 \frac{\partial \sigma}{\partial t} + a_0 \sigma = b_2 \frac{\partial^2 \varepsilon}{\partial t^2} + b_1 \frac{\partial \varepsilon}{\partial t} + b_0 \varepsilon \qquad (34)$$

where

$$a_2 = 1, \quad a_1 = \left(\frac{1}{\tau_1} + \frac{1}{\tau_2} \right), \qquad\qquad a_0 = \frac{1}{\tau_1 \tau_2}$$

$$b_2 = k_g, \quad b_1 = \left[\frac{1}{\tau_1} (k_e + k_1) + \frac{1}{\tau_2} (k_e + k_2) \right], \quad b_0 = \frac{k_e}{\tau_1 \tau_2} \qquad (35)$$

Many authors employ the differential equation formulation written in the operator form, $U\sigma = Q\varepsilon$, where U and Q are the differential operators

$$U = a_n \frac{\partial^n}{\partial t^n} + \ldots + a_1 \frac{\partial}{\partial t} + a_0$$

$$Q = b_m \frac{\partial^m}{\partial t^m} + \ldots + b_1 \frac{\partial}{\partial t} + b_0 \qquad (36)$$

Letting L denote the Laplace transform, we obtain

$$\Re(s) = L\{Q / U\} = \frac{b_m s^m + \ldots + b_1 s + b_0}{a_n s^n + \ldots + a_1 s + a_0} \qquad (37)$$

4.2 *The Boltzmann superposition integral*

Recall that the transformed relaxation modulus is related simply to the associated viscoelastic modulus in the Laplace plane as

$$\bar{E}(s) = \frac{1}{s} \Re(s)$$

Since $s\bar{x} = \dot{\bar{x}}$, the following relations hold

$$\bar{\sigma}(s) = \Re(s)\bar{\varepsilon}(s) = s\bar{E}(s)\bar{\varepsilon}(s) = \dot{\bar{E}}(s)\bar{\varepsilon}(s) = \bar{E}(s)\dot{\bar{\varepsilon}}(s)$$

Inverting $\dot{\bar{E}}(s)\bar{\varepsilon}(s)$ and $\bar{E}(s)\dot{\bar{\varepsilon}}(s)$ leads (see Tschoegl 1989) to the four equivalent convolution integrals

$$\sigma(t) = \int_0^t E(t-\lambda)\dot{\varepsilon}(\lambda)\mathrm{d}\lambda$$

$$= -\int_0^t E(\lambda)\dot{\varepsilon}(t-\lambda)\mathrm{d}\lambda \qquad\qquad (38)$$

$$= E_g\varepsilon(t) - \int_0^t \dot{E}(t-\lambda)\varepsilon(\lambda)\mathrm{d}\lambda$$

$$= E_g\varepsilon(t) + \int_0^t \dot{E}(\lambda)\varepsilon(t-\lambda)\mathrm{d}\lambda$$

where $G(t = 0) = G_g$. These relations are forms of Duhamel's formula and are also referred to as Boltzmann superposition integrals, or as hereditary integrals. $E(t)$ can thus be interpreted as the stress, $\sigma(t)$, resulting from an input of unit constant strain.

If stress rather than strain is the input, an analogous development, for the first equation, leads to

$$\varepsilon(t) = \int_0^t D(t-\lambda)\dot{\sigma}(\lambda)\mathrm{d}\lambda \qquad\qquad (39)$$

where, $D(t)$, the strain response to an input of unit constant stress, is the quantity defined earlier as the creep compliance. In the Laplace transform plane the creep compliance and the relaxation modulus are related by the simple formula

$$\bar{\sigma}(s) = s\bar{E}(s)\bar{\varepsilon}(s)$$
$$\bar{\varepsilon}(s) = s\bar{D}(s)\bar{\sigma}(s) \qquad\qquad (40)$$
$$\bar{\sigma}(s)\bar{\varepsilon}(s) = s^2\bar{E}(s)\bar{D}(s)\bar{\sigma}(s)\bar{\varepsilon}(s) \quad\rightarrow\quad \bar{E}(s)\bar{D}(s) = 1/s^2$$

and therefore

$$\int_0^t E(t-\lambda)D(\lambda)\mathrm{d}\lambda = \int_0^t E(\lambda)D(t-\lambda)\mathrm{d}\lambda = t \qquad\qquad (41)$$

Thus $s\bar{E}(s)$ and $s\bar{D}(s)$ and are simply each others reciprocal in the complex Laplace transform plane, but are linked through a convolution integral on the real time axis. To obtain the creep compliance from the relaxation modulus, or vice versa, one must therefore solve an integral equation. Methods for doing this exist (Hopkins-Hamming, see Tschoegl 1989). The Emri-Tschoegl (1992) algorithm referred to above offers a particularly simple means for obtaining a discrete distribution of retardation times from a distribution of relaxation times, and vice versa.

The Boltzmann integral is the flip side of the operator equation, both assuming the same form in the Laplace plane. Also, both may be seen to be a natural consequence of our earlier development in terms of linear mechanical models. The applicability of either the operator equation or the Boltzmann integral to experimental data provides a definition of linearity which is more stringent than the ability to superimpose a series of relaxation curves through normalization by the applied strain. This ability is referred to as *stress-strain linearity*. In addition, however, there is also time-dependence linearity, also called *time-shift invariance*. This requires that the time dependence of the mechanical properties be describable by linear differential equations (see the operator equation), or, equivalently, by the Boltzmann superposition integral. It is possible for data to pass the multiplicative test implied by the ability to be normalized into a single relaxation curve, but still fail the more stringent additivity test implied in the Boltzmann integral.

5 RELAXATION AND RETARDATION SPECTRA

Previously, we have used combinations of discrete spring-dashpot elements to model the distribution of relaxation times inherent in even 'single' polymer transitions. These discrete models are sometimes awkward, and one occasionally finds it convenient to pass to the limit so as to develop continuous functions. These are the relaxation spectrum, $H(\tau)$, and the retardation spectrum, $L(\tau)$. In terms of these the function introduced earlier becomes

$$E(t) = k_e + \int_{-\infty}^{\infty} H(\tau) \exp(-t/\tau) \, d\ln \tau$$
$$= k_g - \int_{-\infty}^{\infty} H(\tau) [1 - \exp(-t/\tau)] \, d\ln \tau \tag{42}$$

$$E'(\omega) = k_e + \int_{-\infty}^{\infty} H(\tau) \frac{\omega^2 \tau^2}{1 + \omega^2 \tau^2} \, d\ln \tau$$
$$= k_g - \int_{-\infty}^{\infty} H(\tau) \frac{1}{1 + \omega^2 \tau^2} \, d\ln \tau \tag{43}$$

$$E''(\omega) = \int_{-\infty}^{\infty} H(\tau) \frac{\omega \tau}{1 + \omega^2 \tau^2} \, d\ln \tau \tag{44}$$

and

$$D(t) = \frac{1}{k_g} + \int_{-\infty}^{\infty} L(\tau) [1 - \exp(-t/\tau)] \, d\ln \tau$$
$$= \frac{1}{k_e} - \int_{-\infty}^{\infty} L(\tau) \exp(-t/\tau) \, d\ln \tau \tag{45}$$

$$D'(\omega) = \frac{1}{k_g} + \int_{-\infty}^{\infty} L(\tau) \frac{1}{1+\omega^2\tau^2} \, d\ln\tau$$

$$= \frac{1}{k_e} - \int_{-\infty}^{\infty} L(\tau) \frac{\omega^2\tau^2}{1+\omega^2\tau^2} \, d\ln\tau \tag{46}$$

$$D''(\omega) = \int_{-\infty}^{\infty} L(\tau) \frac{\omega\tau}{1+\omega^2\tau^2} \, d\ln\tau \tag{47}$$

Many researchers feel that the continuous distributions $H(\tau)$ or $L(\tau)$ in Equations (42) to (47) are more convenient than the representations in terms of spring-dashpot analogies, and a good deal of the polymer science litera-ture is devoted to the extraction of these functions from various forms of experimental data. Several techniques have been employed for this purpose. One of these, which will be described here for illustration in spite of its usually poor accuracy, is called *Alfrey's approximation*. From Equation (42), we can write

$$E(t) - k_e = \int_{-\infty}^{\infty} H(\tau) \exp(-t/\tau) \, d\ln\tau$$

Differentiating with respect to $\ln\tau$ gives

$$\frac{d[E(t)]}{d\ln\tau} = \int_{-\infty}^{\infty} H(\tau) \left[-\frac{t}{\tau} \exp(-t/\tau) \right] d\ln\tau \tag{48}$$

The function in brackets can be approximated as a 'filter' which has the value $-\exp(-1)$ at $t = \tau$ and zero elsewhere. Assuming $H(\tau)$ is constant in this 'filter interval', one obtains

$$H(\tau) = -\left[\frac{dE(t)}{d\ln t} \right]_{t=\tau} \tag{49}$$

so that $H(\tau)$ can be found from the slope of the relaxation modulus on logarithmic coordinates. Extensive discussions of the means available to obtain approximation to the continuous spectra can be found in the texts by Ferry (1980), and by Tschoegl (1989). More recent developments using discrete spectra obviate the use of these approximations. These develop-ments are the subject of the next section.

6 DISCRETE SPECTRAL DISTRIBUTIONS

The relaxation and retardation functions given by Equations (42) through (47) can be expressed in terms of discrete spectral distributions as

$$E(t) = k_e + \sum_{i=1}^{\infty} H_i \exp\left(-t/\tau_i\right) = k_g - \sum_{i=1}^{\infty} H_i[1 - \exp\left(-t/\tau_i\right)] \quad (50)$$

$$E'(\omega) = k_e + \sum_{i=1}^{\infty} H_i \frac{\omega^2 \tau_i^2}{1+\omega^2\tau_i^2} = k_g - \sum_{i=1}^{\infty} H_i \frac{1}{1+\omega^2\tau_i^2} \quad (51)$$

$$E''(\omega) = \sum_{i=1}^{\infty} H_i \frac{\omega\tau_i}{1+\omega^2\tau_i^2} \quad (52)$$

$$D(t) = \frac{1}{k_g} + \sum_{i=1}^{\infty} L_i \left[1 - \exp\left(-t/\tau_i\right)\right] = \frac{1}{k_e} - \sum_{i=1}^{\infty} L_i \exp\left(-t/\tau_i\right) \quad (53)$$

$$D'(\omega) = \frac{1}{k_g} + \sum_{i=1}^{\infty} L_i \frac{1}{1+\omega^2\tau_i^2} = \frac{1}{k_e} - \sum_{i=1}^{\infty} L_i \frac{\omega^2\tau_i^2}{1+\omega^2\tau_i^2} \quad (54)$$

and

$$D''(\omega) = \sum_{i=1}^{\infty} L_i \frac{\omega\tau_i}{1+\omega^2\tau_i^2} \quad (55)$$

In these equations the sets $\{H_i, \tau_i\}$, and $\{L_i, \tau_i\}$, where $i = 1, 2, ..., N$ represent the discrete distributions of relaxation and retardation times.

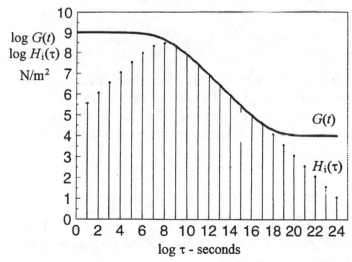

Figure 12. Prediction of a discrete spectrum from the shear relaxation modulus using the Emri-Tschoegl algorithm.

Recently several recursive computer algorithms were proposed (Baumgaertel & Winter 1989, 1992; Honercamp 1989; Honercamp & Weese 1989, 1993; Elster et al. 1991; Carrot et al. 1992; Simhambhatla & Leonov 1993) that generate such discrete distributions. These represent, in effect, the terms in Wiechert or Kelvin models. Probably the most convenient algorithm for determining discrete spectra, equally spaced on the log τ axis, is the algorithm of Emri & Tschoegl (1993, 1994). This algorithm differs from all others referenced above by scanning the data through 'windows' straddling each consecutive relaxation or retardation time over a span of two logarithmic decades. This 'windowing' approach circumvents the ill-posedness inherent in solving the Fredholm integral equations of the first kind, Equations (42) to (47), when an attempt is made to utilize the entire data. Consecutive spectrum lines are evaluated through iteration starting with the longest relaxation or retardation time. The iteration is abandoned when the difference between two consecutive iterations is less than a preset criterion.

Figure 12 shows a synthetic relaxation modulus, $G(t)$, shown by the solid line, and the discrete spectrum, $H_i(\tau)$, obtained from it, represented by the vertical lines.

7 EFFECT OF TEMPERATURE, PRESSURE, AND MOISTURE

Temperature, pressure, moisture and other diluents such as plasticizers have significant effects on the material properties of a composite material, particularly on the matrix component. The effect of temperature is best understood and will be discussed first.

7.1 *The effect of temperature*

An increase in temperature speeds both relaxation and retardation. For most amorphous polymers (excluding block and graft copolymers but including random copolymers) a change in temperature equally affects all relaxation (or retardation) times responsible for the *main or primary transition*. If so, then

$$\frac{\tau_i(T)}{\tau_i(T_{\text{ref}})} = a_{T_{\text{ref}}}(T) = a_T \tag{56}$$

i.e. the ratio between two relaxation times, one at the temperature T, the other at a reference temperature $T_{\text{ref}} > T$ is same for *all* relaxation times. This is illustrated in Figure 13.

The ratio is commonly abbreviated to a_T but it must be remembered that

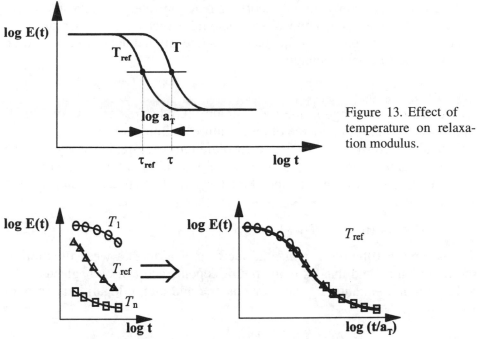

Figure 13. Effect of temperature on relaxation modulus.

Figure 14. Shifting isothermal data segments ($T_1 < T_{ref} < T_n$) to obtain a mastercurve.

a_T a function of T and that the form of this *temperature function* is different for each reference temperature.

It is clearly also subject to the validity of the assumption embodied in Equation (56) above. This assumption is reasonable for the classes of polymers mentioned. It follows then that data taken isothermally at different constant temperatures and plotted against log t can be translated along the logarithmic time axis to form a *mastercurve* as a function of log (t/a_T) valid at T_{ref}, as shown in Figure 14.

Mastercurves obtained from frequency-dependent functions must, of course, be plotted against log (ωa_T). A material that obeys Equation (56) is called *thermorheologically simple*. It is important to note that apparent superposability of isothermal temperature segments does not necessarily imply thermorheological simplicity (Fesko & Tschoegl 1971).

When Equation (56) does apply, the shifting procedure furnishes a set of discrete values of the temperature function. One wishes, however, to possess this function in an analytic form. Relaxation (or retardation) may be regarded as a physical process during which the material is carried from state A into state B. The relaxation or retardation time is the time constant for this process. It is the reciprocal of the rate at which the physical change takes place and is thus akin to the reciprocal of a chemical reaction rate.

Provided that the conventional theory of reaction rates is applicable to re-laxation-retardation phenomena, one should expect the temperature dependence of these processes, in analogy to that of chemical reactions, to be given by the Arrhenius equation

$$\tau = \alpha \exp(\Delta G / RT) \tag{57}$$

where ΔG is the change in the free enthalpy (Gibbs free energy) of activation for the process per degree of temperature and mole of material, R is the universal gas constant, and T is the absolute temperature. Discussion of the thermodynamic or molecular nature of the activation energy is beyond our scope here. At a different temperature (say, at the reference temperature, T_{ref}) we have

$$\tau' = \alpha \exp(\Delta G / RT_{ref}) \tag{58}$$

We assume that the pre-exponential factor α is at most a weak function of the temperature and that its temperature dependence can be neglected. Dividing the last equation by the previous one and taking logarithms on both sides we find

$$\log \tau / \tau' = \frac{\Delta G}{2.303R} \left(\frac{1}{T} - \frac{1}{T_{ref}} \right) = \log a_T \tag{59}$$

where we have used Equation (56) to introduce a_T. A material whose responses display temperature dependence according to Equation (59) is said to show *Arrhenius behaviour*. The temperature dependence of the so-called *secondary transitions* of many polymeric materials can be described adequately by this equation. It fails, however, for the primary or main transitions in such materials. The reason for this is easily seen. By Equation (59) the relaxation time becomes infinite at the absolute temperature $T = 0$. An infinite relaxation time implies that the material does not relax, i.e. it is purely elastic in its response. In many materials relaxation (or retardation) virtually ceases well above $T = 0$. Thus the segmental motion of polymer chains that gives rise to the phenomena of relaxation or retardation effectively ceases when the temperature is lowered below the value referred to as the *glass transition temperature*, T_g. Gibbs & DiMarzio (see Ferry 1980) have predicted the existence in polymers of a second order thermodynamic transition temperature, denoted T_L, that could be reached only by an infinitely slow cooling process. The experimentally determined glass transition temperature, T_g, is a kinetic phenomenon. When measured at the usual experimental cooling rate of about 5-15°C/min, it is about 50° above T_L, i.e. we can set $T_g \cong T_L + 50$. Using this relation we may modify Equation (58) to allow it to become infinite when $T = T_L$. We have

$$\tau = \alpha \exp\left[\frac{\Delta G}{R(T - T_L)}\right] \tag{60}$$

T_L is thus the threshold temperature (Latin *Limen*, threshold) below which the relaxation or retardation process cannot be activated thermally. Equation (60) comprises the Arrhenius equation as the special case when $T = 0$. We now have

$$\log a_T = \frac{\Delta G}{2.303R}\left(\frac{1}{T - T_L} - \frac{1}{T_{ref} - T_L}\right) \tag{61}$$

which may be rearranged to

$$\log a_T = -\frac{c_1^r(T - T_{ref})}{c_2^r + T - T_{ref}} \tag{62}$$

where

$$c_1^r = -\frac{\Delta G}{2.303Rc_2^r} \tag{63}$$

and

$$c_2^r = T_{ref} - T_L \tag{64}$$

are constants depending on the reference temperature chosen, apart from T_L and the activation energy ΔG. Equation (62) is known as the WLF-equation after Williams, Landel and Ferry, who introduced it in this form (see Ferry 19980) to describe the temperature dependence of polymeric materials. The WLF-equation if found to be applicable to the majority of glass forming substances, including inorganic glasses (e.g. sulfur, silicates, etc.), organic glasses (e.g. glucose and glycerol), and amorphous metals (e.g. iron).

Of the constants, c_1^r shows a small, and c_2^r a somewhat larger variation with molecular structure. The values of the constant c_1^l and c_2^l for another reference temperature T_1 can be found from c_1^r, c_2^r and T_{ref} by the symmetrical equations

$$c_1^l - T_1 = c_2^r - T_{ref} \tag{65}$$

and

$$c_1^l c_2^l = c_1^r c_2^r \tag{66}$$

The coefficients c_1 and c_2 have received interpretations in terms of certain molecular parameters. The most widely used of these interpretations is that in terms of the *(fractional) free volume, f.* Any effect that increases f speeds up the relaxation or retardation times. For details the treatise by Ferry (1980) should be consulted.

Because c_1 and c_2 are generally weak functions of the molecular structure, Williams, Landel and Ferry (see Ferry 1980) found that the temperature dependence of the mechanical properties of many polymers can often be expressed with good approximation by a single-parameter form of the WLF-equation in which the constants c_1 and c_2 receive fixed values and the temperature dependence is referred to the glass transition temperature, T_g. Equation (62) then becomes

$$\log a_T = -\frac{17.44\ (T - T_g)}{51.6 + T - T_g} \tag{67}$$

This 'universal' form of the WLF-equation allows $\log a_T$ to be predicted if T_g is known. However, it should only be used as an approximation, and c_1 and c_2, should be determined whenever possible.

Above about $T_g + 100$ experimental shift data appear to deviate from the WLF equation. Applications of Equations (62) and (67) should therefore be restricted to between T_g and $T_g + 100$.

Below the glass transition temperature materials are exposed to *physical ageing* (this phenomenon will be discussed in Section 8). Consequently the WLF equation can be used below T_g only if the material is in thermodynamic equilibrium, i.e. is not undergoing physical ageing.

7.2 *The effect of pressure*

Pressure has an effect opposite to that of temperature. An increase in pressure slows the relaxation or retardation processes. Again, when all relaxation or retardation times are equally affected by a change in pressure, we speak of a *piezorheologically* simple material. Isobaric-isothermal segments recorded under different constant pressures can then be shifted into superposition to yield a mastercurve. The appropriate equation which describes both temperature and pressure dependence is the FMT-equation (Fillers & Tschoegl 1977, Moonan & Tschoegl 1983) which comprises the WLF-equation as a special case. The equation reads

$$\log a_{T,P} = -\frac{c_1^{rr}[T - T_{\text{ref}} - \theta(P)]}{c_2^{rr}(P) + T - T_{\text{ref}} - \theta(P)} \tag{68}$$

where T and P are the (absolute) temperature and pressure, respectively, and $\theta(P)$ is given by

$$\theta(P) = c_3^r(P)\ln\left[\frac{1 + c_4^r P}{1 + c_4^r P_{\text{ref}}}\right] - c_5^r(P)\ln\left[\frac{1 + c_6^r P}{1 + c_6^r P_{\text{ref}}}\right] \tag{69}$$

Of the two superscripts in c_1^{rr} and $c_2^{rr}(P)$, the first refers to the reference temperature and the second to the reference pressure. All parameters are dependent on the choice of the reference temperature, c_1^{rr} and $c_2^{rr}(P)$ are dependent on the choice of reference pressure also, and c_2^{rr}, c_3^{r} and c_5^{r} depend, in addition, on the experimental pressure, P. When $P = P_{ref}$, $\theta(P) = 0$, and the FMT equation reduces to the WLF equation. All parameters can be determined experimentally. For details and for the interpretation of the parameters in molecular terms, the paper by Moonan & Tschoegl (1983) should be consulted.

7.3 *The effect of moisture and other diluents*

Moisture and other diluents (plasticizers), molecular weight (free chain ends), and strain all can increase the free volume and, hence, increase the relaxation or retardation times of a polymeric material (Cardon et al. 1986, Ferry 1980).

8 PHYSICAL AGEING

An amorphous polymeric material in the glassy state, i.e. below its glass transition temperature, is generally not in thermodynamic equilibrium because molecular rearrangements initiated during the manufacturing process may not have ceased (Kovacs 1964, Struik 1978, Sullivan 1990). The phenomenon, known as *physical ageing*, results from the rapid changes in temperature and pressure to which the material had been exposed during, and particularly at the end, of the manufacturing process (as when an injection molded part emerges from the die). Physical ageing sets in when the rate at which the temperature and/or pressure changes is faster than the characteristic rate of the molecular rearrangements. These rearrangements are accompanied by changes in volume, as well as in mechanical and other physical properties. The volumetric changes are usually small and may be negligible in less demanding situations. On the other hand the variations of the mechanical, electrical and optical properties can be of several orders of magnitude, and should therefore be taken into account in most engineering applications.

A material exhibiting physical ageing attains its final volume slowly. Figure 15 shows the typical volume retardation response of poly(vinyl acetate) subjected to temperature jumps from well above T_g to the temperatures indicated in the figure (Kovacs 1964). These volume contractions were measured in the isothermal state, i.e. after constant temperature had been reached throughout the entire volume of the specimen. Note that for larger

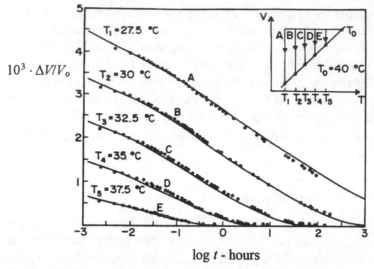

Figure 15. Volume retardation exhibited by physical ageing of poly(vinyl acetate).

temperature jumps the volume retardation process can proceed over several months or even several years.

When a polymeric material is exposed to a more complex temperature and/or pressure history the resulting response can be rather unexpected. An increase in the temperature of the material can, for example, cause the volume to *contract*, contrary to expectation. It is one of the serious consequences of physical ageing that contraction of the matrix can substantially contribute to stress concentration around the fibers and thus influence the durability of the composite material. The solid lines are obtained from a theory describing volume retardation developed by Knauss & Emri (1981, 1987) (Section 10.5).

9 MULTIAXIAL STRESS-STRAIN STATES

All of the previous descriptions have been based on the assumption of a simple stress state in which a specimen is subjected to uniaxial tension. This loading is germane to laboratory characterization tests, but the information obtained from such tests must be cast in terms of general stress and strain states that are described by the (symmetric) stress and strain tensors, σ_{ij} and ε_{ij}. The extension to multiaxial stress states is usually achieved by noting that the molecular conformational rearrangements which engender viscoelastic relaxation are driven primarily by the shearing components of the applied stress; the hydrostatic component of stress gives rise to a much lower order of deformation. It the deformation is small (theoretically infinitesimal), deformation of shape (shear) and deformation of size (bulk) can be neatly separated. Accordingly, one seeks to dissociate a given stress state into its dilatational (isotropic, or hydrostatic) and deviatoric (shearing) components.

Using index notation the deviatoric components of the (symmetric) stress and strain tensors become

$$S_{ij} = \sigma_{ij} - \frac{1}{3}\sigma_{kk}\delta_{ij} \tag{70}$$

$$e_{ij} = \varepsilon_{ij} - \frac{1}{3}\varepsilon_{kk}\delta_{ij} \tag{71}$$

where δ_{ij} is the unit tensor, and the σ_{kk} and ε_{kk} are the traces, i.e. the sums of the components on the main diagonal.

The constitutive relations for an isotropic elastic solid can be written as

$$S_{ij} = 2Ge_{ij} \tag{72}$$

and

$$\sigma_{kk} = 3K\varepsilon_{kk} \tag{73}$$

where the factor of two appears in Equation (72) in order to account for the tensor definition of shear strain. This is half of the 'classical' definition for which handbook values of the shear modulus have been tabulated. K is the bulk modulus, defined as

$$K = \frac{-p}{\Delta V / V} = \frac{1}{3}\frac{\sigma_{KK}}{\varepsilon_{KK}} \tag{73}$$

The shear modulus, G, and the bulk modulus, K, are related to the Young's modulus, E, and Poisson's ratio, v, by

$$E = \frac{9GK}{3K + G} \tag{75}$$

$$v = \frac{3K - 2G}{6K + 2G} \tag{76}$$

The viscoelastic analogs of Equations (72) and (73) then become

$$\bar{S}_{ij}(s) = 2s\bar{G}(s)\bar{e}_{ij}(s) \tag{77}$$

$$\bar{\sigma}_{kk}(s) = 3s\bar{K}(s)\bar{\varepsilon}_{kk}(s) \tag{78}$$

and these viscoelastic operators may be related to the tensile operators by

$$\overline{E}(s) = \frac{9\overline{G}(s)\overline{K}(s)}{3\overline{K}(s) + \overline{G}(s)} \tag{79}$$

$$\overline{v}(s) = \frac{3\overline{K}(s) - 2\overline{G}(s)}{2s[3\overline{K}(s) + \overline{G}(s)]} \tag{80}$$

Proper characterization of composite materials requires knowledge of either the time-dependent bulk modulus, $K(t)$, or the time-dependent Poison's ratio, $v(t)$. These are difficult to determine experimentally. As shown schematically in Figure 16 below, $K(t)$ is orders of magnitude larger than $G(t)$ but experiences much smaller relaxation, and so does the closely related $v(t)$. In the transform plane $\overline{K}(s)$ and $\overline{v}(s)$ are given by

$$\overline{K}(s) = \frac{\overline{E}(s)\overline{G}(s)}{9\overline{G}(s) - 3\overline{E}(s)} \tag{81}$$

and

$$\overline{v}(s) = \frac{\overline{E}(s) - 2\overline{G}(s)}{2s\overline{G}(s)} \tag{82}$$

It should therefore be possible in principle to find either the time-dependent bulk modulus or the time-dependent Poisson's ratio from the much more easily determined $E(T)$ and $G(t)$ if it were not for the tremendous and taxing accuracy required. In practice one therefore commonly resorts to taking $K(t) = K_e$ to be finite but constant, deeming only the shear response to be viscoelastic. In that case $s\overline{K}(s) = K_e$ and we have

$$\overline{G}(s) = \frac{3K_e\overline{E}(s)}{9K_e - s\overline{E}(s)} \tag{83}$$

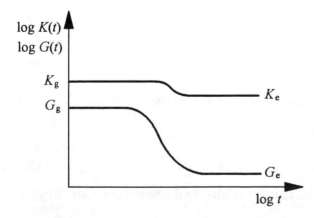

Figure 16. Schematic representation of shear and bulk modulus in logarithmic coordinates.

Secondly, if K is assumed to be not only constant but infinite (i.e. the material is considered to be effectively incompressible), then

$$\overline{G}(s) = \frac{1}{3}\,\overline{E}(s) \tag{84}$$

and

$$s\overline{v}(s) = v = \frac{1}{2} \tag{85}$$

In the case of material isotropy (properties not dependent on direction of measurement), at most two viscoelastic operators – say $\overline{G}(s)$ and $\overline{K}(s)$ – will be necessary for the full characterization of a material. For materials exhibiting lower orders of symmetry, such as the transversely isotropic material which is obtained from drawing processes, more descriptors will be necessary. As is demonstrated in standard texts in solid mechanics, a transversely isotropic material requires five constitutive descriptors, an orthotropic material requires nine, and a triclinic material twenty-one. If the material is both viscoelastic and anisotropic, these are the number of viscoelastic functions that will be required. Not many researchers have been interested in grinding through such tedious manipulations. However, several of the large general-purpose finite-element computer codes do have the capability for these analyses.

10 TIME-DEPENDENT NON-LINEAR BEHAVIOUR

Essentially all composite materials exhibit nonlinearity at sufficiently large strains. Some researchers hold that true linearity as demanded by the Boltzmann integral is never achieved by any known material even at extremely low strains. The extension of the theory to incorporate non-linear effects is fraught with difficulties, and as yet there is no general consensus on how best to proceed. The literature contains numerous attempts to develop non-linear predictions. These are virtually always derived from a single type of experimental test. This is usually the tensile test because it is relatively easy to perform and analyze. Unfortunately, most practical applications involve the multiple stress-strain states discussed in Section 9. The currently available predictions are therefore essentially curve-fitting approaches. Such approaches are useful if the results are employed for interpolation within the domain of validity of the procedure but great caution must be exercised in rewriting them for modes of deformation other than uniaxial stress.

To show that the prediction is *constitutive*, i.e. that the material parameters and functions it contains truly describe the *constitution* of the matter of which the material consists, it is necessary to verify that behaviour in different modes of deformation can be described with the appropriately adapted formalism using the same material parameters and functions. Very few (if any) such verification has been produced so far. Constitutiveness can never be proved but it is relatively simple to disprove it. Apart from uniaxial tension any non-linear procedure should be tested in at least equibiaxial tension as well.

We concentrate here on *time-dependent non-linear methods* and refer the reader to any of many text on elasto-plastic non-linear methods such as the theories of Rivlin & Green, Metzner & White, and Bernstein, Kearsley & Zapas (BKZ).

10.1 *Model based on time shift invariance*

Chang et al. (1971) have found that time shift invariance is often preserved at moderately large strains. All that is required is to introduce an appropriate non-linear expression for the strain in the Boltzmann superposition integral. This model is readily adaptable to different modes of deformation and should be constitutive as long as the non-linear strain expression itself is constitutive. The domain of validity of this model must be established for each material because it cannot be predicted theoretically. When applicable, this model appears to be by far the simplest for describing non-linear time-dependent behaviour. It apparently has not yet been tested on composite materials.

10.2 *Non-linear spring-and-dashpot models*

One way of dealing with time-dependent nonlinearity is to introduce non-linear dashpots and non-linear springs into the Wiechert or Kelvin models.

This approach is limited to simple (usually uniaxial) stress-strain conditions, and suffers from the fact that the behaviour of models, constituted from several such units, cannot easily be expressed in mathematical terms.

10.3 *Multiple Boltzmann Integrals*

One of the earliest more general approaches, advocated by Schapery

$\sigma = k \exp t$ $\dot{\varepsilon} = A \sinh(\alpha \sigma)$ Figure 17. Non-linear spring and dashpot.

(1961), relates the strain to the stress via generalized (multiple) Boltzmann integrals. The non-linear material response is expressed by a series of creep compliance functions, $D_1(t)$, $D_2(t)$, $D_3(t)$, ... This gives

$$\varepsilon(t) = \int_0^t D_1(t-\lambda) \frac{\partial\sigma(\lambda)}{\partial\lambda} \, d\lambda$$

$$+ \int_0^t \int_0^t D_2(t-\lambda_1, t-\lambda_2) \frac{\partial\sigma(\lambda_1)}{\partial\lambda_1} \frac{\partial\sigma(\lambda_2)}{\partial\lambda_2} \, d\lambda_1 d\lambda_2$$

$$+ \int_0^t \int_0^t \int_0^t D_3(t-\lambda_1, t-\lambda_2, t-\lambda_3) \tag{86}$$

$$\frac{\partial\sigma(\lambda_1)}{\partial\lambda_1} \frac{\partial\sigma(\lambda_2)}{\partial\lambda_2} \frac{\partial\sigma(\lambda_3)}{\partial\lambda_3} \, d\lambda_1 d\lambda_2 d\lambda_3$$

$$+ \ldots$$

This approach, while mathematically unassailable, suffers from its considerable intractability in design applications, from the difficulty of selecting the memory functions in terms of experimental data, and most importantly, lack of physical meaning of the various terms. Besides, the infinite series must, in practice be replaced with no more than two to three terms. The truncated series is, in addition, unlikely to describe behaviour in different modes of deformation with the same compliance functions.

10.4 *Non-linear single integral*

Another approach, originally also proposed by Schapery and later elaborated by Brüller (1987), introduces non-linear functions into the Boltzmann integral equation,

$$\varepsilon(t) = g_0(\sigma)D_0\sigma + g_1(\sigma) \int_0^t \Delta D(t-\lambda) \frac{\partial[g_2(\sigma)\sigma(\lambda)]}{\partial\lambda} \, d\lambda \tag{87}$$

The nonlinearity is brought in through the g_1, g_2 and g_3's that are functions of the applied time-dependent stress.

10.5 *The internal time model*

Knauss & Emri (1981, 1987) have proposed a non-linear thermoviscoelastic model based on the free volume concept that has been shown to be an appropriate concept for modeling the influence of temperature, pressure, and mechanical loading. Here

$$S_{ij}(t) = 2 \int_0^t G[t'(t) - \lambda'(t)] \frac{\partial e_{ij}(\lambda)}{\partial\lambda} \, d\lambda$$

and   (88)

$$\sigma_{kk} = 3 \int_0^t K[t'(t) - \lambda'(t)] \frac{\partial \theta(\lambda)}{\partial \lambda} d\lambda$$

The model encompasses the time dependent nonlinearity through the material's *internal time* which is the measure of the rate of molecular rearrangements. An increase in temperature, for example, will increase the rate of molecular rearrangements and consequently 'speed up' the internal time. Since the molecular dynamics can be influenced by either the temperature, the pressure, the moisture and/or the mechanical loading, the internal time should be a functional of the three physical quantities. This relation can be expressed via the free volume, by

$$t'(t) - \lambda'(t) = \int_\lambda^t \frac{d\xi}{\Phi[T(\xi), \theta(\xi), c(\xi)]}$$ (89)

$$\log \Phi[T(\xi), \theta(\xi), c(\xi)] = \frac{b}{2.303} \left(\frac{1}{f[T(\xi), \theta(\xi), c(\xi)]} - \frac{1}{f_o} \right)$$ (90)

and

$$f[T(\xi), \theta(\xi), c(\xi)] = f_o + f_T + f_\theta + f_c$$ (91)

Here f_o is the fractional free volume in the equilibrium state, while the f_T, f_θ and f_c are changes of the free volume caused by temperature, mechanical loading, and moisture. These contributions are given by

$$f_T = \int_0^t \alpha(t - \lambda) \frac{\partial T(\lambda)}{\partial \lambda} d\lambda$$ (92)

$$f_\theta = \theta = \frac{1}{3} \int_0^t M(t - \lambda) \frac{\partial \sigma_{kk}(\lambda)}{\sigma \lambda} d\lambda$$ (93)

and

$$f_c = \int_0^t \gamma(t - \lambda) \frac{\partial c(\lambda)}{\partial \lambda} d\lambda$$ (94)

Here $\alpha(t)$, $M(t)$, and $\gamma(t)$ are the *thermal creep* function, the *bulk creep compliance*, and the solvent-related *volume creep function*. All material functions used in the model are *linearly viscoelastic*.The time-dependent nonlinearity (to which the model is limited) results from the introduction of the non-linear link between the experimental time and the material's internal time (see Equation (89)).

An illustration of the applicability of the model is shown in Figure 15, which presents physical ageing of poly(vinyl acetate) subjected to various temperature jumps. The analytical prediction, represented by the solid lines, is compared with the experimental data obtained by Kovacs (1964), shown by squares.

11 VISCOELASTIC STRESS ANALYSIS

Material behaviour is characterized by *constitutive equations*. A simple such equation for a purely elastic material is $\sigma = G\varepsilon$. This equation links the shear stress, σ, with the shear strain, ε, through the shear modulus, G. The material behaviour characterized by this equation can be dealt with in two ways: 1) we may be interested in a complicated stress-strain relation but take the material parameter, i.e. the modulus, G, as known; this approach is that of stress analysis; 2) we may be interested in determining the material parameter, here, G. This requires a simple stress-strain relation and belongs to the realm of *rheology*. In the following we are going to consider stress analysis applied to viscoelastic materials. The above simple constitutive relation then becomes

$$\sigma(t) = \int_0^{\varepsilon(t)} G(t - u)\,d\varepsilon(u)$$

a relation between the time dependent quantities.

11.1 *Elastic stress analysis*

Before turning to viscoelastic stress analysis we consider a general solid mechanics problem in which a body is subjected to interior body forces f_i, applied tractions $\hat{T}_i$ over a portion of its surface S_t, and imposed displacements $\hat{u}_i$ over the remainder of its surface S_u, as shown in Figure 18.

The field equations relating the six components of the (symmetric) stress tensor, $\sigma_{ij}(x_k)$, the strain tensor, $\varepsilon_{ij}(x_k)$, and the three displacements vectors, $u_i(x_k)$, within the body are:
Three equilibrium equations,

$$\sigma_{ij,j}(x_k) = f_i(x_k) \tag{95}$$

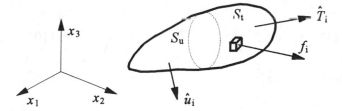

Figure 18. Solid body subjected to internal body forces and external loading.

six equations relating strain and displacement,

$$\varepsilon_{ij}(x_k) = \frac{1}{2}[u_{i,j}(x_k) + u_{j,i}(x_k)] \tag{96}$$

and *six constitutive equations relating stress and strain,*

$$\varepsilon_{ij}(x_k) = \frac{1+v}{E}\sigma_{ij}(x_k) - \frac{v}{E}\delta_{ij}\sigma_{kk}(x_k) \tag{97}$$

In Equations (95) and (96) the subscript after the comma denotes partial differentiation.

We now seek to find the fifteen unknown stress, strain and displacement functions which satisfy these fifteen equations, and in addition satisfy the boundary conditions:

$$\sigma_{ij}(x_k)n_j = \hat{T}_i(x_k) \text{ on } S_t \tag{98}$$

$$u_i(x_k) = \hat{u}_i(x_k) \text{ on } S_u \tag{99}$$

where n_j is the unit normal vector at the surface. In elastic materials, $f_i, \hat{T}_i,$ and $\hat{u}_i$, and may depend on time as well as position without affecting the solution. Time is carried only as a parameter, since no time derivatives appear in the governing equations.

With viscoelastic materials, Equation (97) is replaced by a time-differential equation, which of course complicates the subsequent solution. In many cases, however, the field equations possess certain mathematical properties that permit a solution to be obtained relatively easily.

11.2 *Integral transform solutions*

Turning to the Laplace transformation to simplify dealing with the time dependence encountered in viscoelastic problems, the transformed field equations become

$$\overline{\sigma}_{ij,j}(x_k, s) = \overline{f}_i(x_k, s) \tag{100}$$

$$\overline{\varepsilon}_{ij}(x_k, s) = \frac{1}{2}[\overline{u}_{i,j}(x_k, s) = \overline{u}_{j,i}(x_k, s)] \tag{101}$$

$$\overline{S}_{ij}(x_k, s) = 2\overline{G}(s)\overline{e}_{ij}(x_k, s) \tag{102}$$

$$\overline{\sigma}_{kk}(x_k, s) = 3\overline{K}(s)\overline{\varepsilon}_{kk}(x_k, s) \tag{103}$$

$$\bar{\hat{T}}_i(x_k, s) = \bar{\sigma}_{ij}(x_k, s)n_j \quad \text{on} \quad S_t \tag{104}$$

and

$$\bar{\hat{u}}_i(x_k, s) = \bar{u}_i(x_k, s) \quad \text{on} \quad S_u \tag{105}$$

Note that Equation (97) has been recast into octahedral form yielding two separate equations, Equations (102) and (103), governing changes in geometry and volume, respectively. These equations can be interpreted as representing a stress analysis problem for an elastic body of the same shape as the viscoelastic body, where s is carried as a parameter. Although the description of the shape of the body is unchanged upon passing to the Laplace plane, the transformed boundary constraints $\bar{f}_i, \bar{\hat{T}}_i$ and $\bar{u}_i$, that can be functions of time, will, in general, exhibit an altered spatial distribution.

A vast library is available for the solution of *elastic* problems. In dealing with many *viscoelastic* problems one may use these solutions invoking the so-called *correspondence principle*. According to this principle one finds the corresponding elastic solution and replaces the elastic quantities by the Laplace transforms of the appropriate viscoelastic ones. Retransformation then yields the viscoelastic solution. The correspondence principle can only be applied if the boundary conditions are not themselves functions of time.

The process of viscoelastic stress analysis can then be carried out in the following steps:

1. Determine the nature of the associated elastic problem. If the spatial distribution of the boundary and body-force conditions is unchanged on transformation then the associated elastic problem appears exactly like the original viscoelastic one. Functions which can be written as separable space and time factors will not change spatially on transformation, i.e.

$$\hat{T}_i(x, t) = f(x)g(t) \quad \rightarrow \quad \bar{\hat{T}}_i(s) = f(x)\bar{g}(s)$$

2. Determine the solution to the associated elastic problem.
3. In the elastic solution replace the material constants, E, v, G and K by their viscoelastic analogs $s\bar{E}(s), s\bar{v}(s), s\bar{G}(s)$ and $s\bar{K}(s)$, or their inverses.
4. Replace the applied boundary and body force constraints by their transformed counterparts

$$\hat{T}_i(x_k, t) \quad \rightarrow \quad \bar{\hat{T}}_i(x_k, s) \quad and \quad \hat{u}_i(x_k, t) \quad \rightarrow \quad \bar{\hat{u}}_i(x_k, s)$$

5. Invert the expression obtained in (4) to obtain the solution to the viscoelastic problem in the time plane.

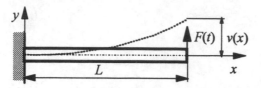

Figure 19. Viscoelastic cantilever beam subjected to a time-dependent load.

11.3 *Example*

As an illustration consider a cantilevered viscoelastic beam subjected to a time-dependent load $F(t) = F_o t$ at its free end, shown in Figure 19.

The associated elastic problem corresponding to this situation appears spatially unchanged, since the space dependence of the applied load is unchanged on transformation. The elastic solution for the stress and deflection of this beam may be found in several handbooks, or may be obtained easily resorting to elementary texts on the mechanics of materials.

For the stress we find

$$\sigma(x, y) = \frac{-F(L-x)y}{I} \tag{106}$$

and for the deflection we have

$$v(x) = \frac{Fx^2}{6EI}(3L-x) = \frac{FDx^2}{6I}(3L-x) \tag{107}$$

where $D = 1/E$.

Note that the solution for the stress does not depend on any of the material properties. Hence the stress in the viscoelastic beam is identical to that in an elastic beam of the same shape. This result will be obtained in any statically determinate problem since the internal stresses in such cases are uniquely determined by means of the equilibrium equations alone.

To determine the viscoelastic displacement, $v(x, t)$, we now replace the elastic constants in Equation (107) by their viscoelastic analogs, and replace the applied time-dependent load by its transform. Since in this case we seek to determine a deformation in terms of a load, it is necessary to replace $s\overline{E}(s)$ by its inverse $s\overline{D}(s)$. This gives

$$\overline{v}(x, s) = \frac{\overline{F}(s)x^2 s\overline{D}(s)}{6I}(3L-x) \tag{108}$$

where

$$\overline{F}(s) = F_o / s^2$$

Let the viscoelastic compliance be represented by a Kelvin model,

$$s\overline{D}(s) = D_g + \sum_i \frac{D_i}{1+\tau_i s} \tag{109}$$

Then Equation (108) becomes

$$\overline{v}(x, s) = \frac{x^2(3L-x)}{6I} F_o \left[\frac{D_g}{s^2} + \sum_i \frac{D_i}{s^2(1+\tau_i s)} \right] \tag{110}$$

The solution in the time plane is now obtained by Laplace inversion as

$$v(x, t) = \frac{x^2(3L-x)}{6I} F_o$$
$$\left\{ D_g t + \sum_i D_i \left[t - \tau_i \left(1 - \exp\left(1 - t/\tau_i\right) \right) \right] \right\} \tag{111}$$

Other examples may be found in the literature on viscoelastic stress analysis.

ACKNOWLEDGEMENTS

The author would like to thank Prof. N.W. Tschoegl for many suggestions and comments that have substantially improved the contents of this paper. The help of Mr R. Cvelbar in preparation of the figures is also appreciated. The financial support by the Slovene Ministry of Science and Technology is gratefully acknowledged.

REFERENCES

Baumgaertel, M. & H.H. Winter 1989. Determination of discrete relaxation and retardation time spectra from dynamic mechanical data. *Rheol. Acta* 28: 511-519.
Baumgaertel, M. & H.H. Winter 1992. Interrelation between discrete and continuous relaxation time spectra. *J. Non-Newtonian Fluid Mech.* 44: 15-30.
Brueller, O.S. 1987. Non-linear characterization of the long term behaviour of polymeric materials. *Polym. Engn. Sci.* :2 56-90.
Cardon, A.H., H.F. Brinson & C.C. Hiel 1989. Non-linear viscoelasticity applied for the study of durability of polymer matrix composites. *Proc. ECCM 3*. Bordeaux.
Cardon, A.H., C.C. Hiel & H.R. Brouwer 1986. Non-linear behaviour of epoxy-matrix composites under combined mechanical and environmental loading. *Proc. Int. Symp. on Comp. Mater.* Beijing.
Carrot, C., J. Guillelt, J. May & J. Puaux 1992. Application of the Marquardt-Levenberg procedure to the determination of discrete relaxation spectra. *Makromol. Chem. Theory Simul.* 1: 215-231.
Chang, W.V., R. Bloch & N.W. Tschoegl 1977. Study of viscoelastic behaviou of un-

crosslinked (gum) rubbers in moderately large deformations. *J. Polym. Sch., Polym. Phys. Ed.* 15: 923-944.

Christensen, R.M. 1979. *Mechanics of composites.* New York: John Wiley.

Elster, C., J. Honercamp & J. Weese 1991. Using regularization methods for the determination of relaxation and retardation spectra of polymeric liquids. *Rheol. Acta*, 31: 161-174.

Emri, I. & N.W. Tschoegl 1993. Generating line spectra from experimental responses. Part I. Relaxation modulus and creep compliance. *Rheol Acta* 33: 31-321.

Emri, I. & N.W. Tschoegl 1994. Generating line spectra from experimental responses. Part IV. Application to experimental data. *Rheol. Acta* 33: 60-70.

Ferry, J.D. 1980. *Viscoelastic properties of polymers.* New York: John Wiley.

Fesko, D.G. & N.W. Tschoegl 1971. Time-temperature superposition in thermorheologically complex materials. *J. Polym. Sci., Part C, Symposia* 35: 56-69.

Fillers, R.W. & N.W. Tschoegl 1977. The effect of pressure on the mechanical properties of polymers. *Trans. Soc. Rheol.* 21: 51-100.

Findlay, W., J.S. Lai & K. Onaran 1976. *Creep and relaxation of non-linear viscoelastic materials.* Amsterdam: North-Holland.

Flugge, W. 1975. *Viscoelasticity.* Berlin: Springer-Verlag.

Honercamp, J. & J. Weese 1989. Determination of the relaxation spectrum by a regularization method. *Macromolecules* 22: 4372-4377.

Honercamp, J. & J. Weese 1993. A non-linear regularization method for the calculation of relaxation spectra. *Rheol. Acta* 32: 65-73.

Knauss, W.G. & I. Emri 1981. Non-linear viscoelasticity based on free volume consideration. *Computer and structures* 13: 123-129.

Knauss, W.G. & I. Emri 1987. Volume change and the nonlinearly thermo-viscoelastic constitution of polymers. *Polym. Eng. Sci.* 27: 86-100.

Kovacs, A.J. 1964. Transition vitreuse dans les polymeres amorphes. Etude phenomenologique. *Adv. Polymer Science* 3: 394-507.

Lou, Y.C. & R.A. Schapery 1971. Viscoelastic characterization of a non-linear fiber reinforced plastics. *J. Comp. Mater.* 5: 208-233.

Moonan, W.K. & N.W. Tschoegl 1983. Effect of pressure on the mechanical properties of polymers. 2. Expansivity and compressibility measurements. *Macromolecules* 16: 55-59.

Schapery, R.A. 1961. On the characterization of non-linear viscoelastic materials. *Polym. Engn. Sci.* 9: 295-310.

Struik, L.E.E. 1978. *Physical ageing in amorphous polymers and other materials.* Amsterdam: Elsevier.

Simhambhatla, M. & A.I. Leonov 1993. The extended Pade-Laplace method for efficient discretization of linear viscoelastic spectra. *Rheol. Acta* 32: 589-600.

Sullivan, J.L. 1990. Creep and physical ageing of composits. *Comp. Sci. Tech.* 39: 207-232.

Tuttle, M.E. & H.F. Brinson 1985. Prediction of the long-term creep compliance of general composition laminates. *Exp. Mech.* 3 89-102.

Tschoegl, N.W. 1989. The phenomenological theory of linear viscoelastic behaviour. Berlin: Springer-Verlag.

Tschoegl, N.W. & I. Emri 1992. Generating line spectra from experimental responses. Part III. Interconversion between relaxation and retardation behaviour. *Intern. J. Poly. Mater.* 18: 117-127.

Damage tolerance and durability of fibrous material systems: A micro-kinetic approach

K. Reifsnider, N. Iyengar, S. Case & Y.L. Xu
Virginia Polytechnic Institute and State University

ABSTRACT: The present paper addresses the question of how to predict the remaining strength and life of fibrous material systems using mechanistic models based on characterizations of behaviour such as creep, creep rupture, and fatigue damage under long-term conditions that may include the effect of aggressive environments. The approach described herein is based on several new concepts developed over the last twelve years, including the 'critical element' concept as a method of correctly setting a continuum mechanics boundary value problem, the 'micro-kinetic' method of representing the rate equations which describe cycle and time-dependent behaviour (or 'evolution'), and micromechanical representations of material principle strengths as a method of integrating the various rate equations, material state information, and local stress details. Examples of application of the philosophy using the MRLife performance simulation code as a collective embodiment of this approach will be presented and compared with observed behaviour of polymer and ceramic composite systems under a variety of loading and environmental conditions.

1 DAMAGE ACCUMULATION: THE PHYSICAL PROBLEM

The details of the present approach are motivated by the physical situation it attempts to represent. The physical situation differs substantially from that found in homogeneous, isotropic materials. Because composites are inhomogeneous (generally made from brittle constituents), and quite often anisotropic, the processes that control their intrinsic strength, and their remaining strength and life under long-term conditions are complex and widely distributed throughout the material system. Matrix cracks, for example, are typically initiated in great numbers (often in patterns that reflect the microstructure (Curtin 1991; Aveston et al. 1971; Reifsnider & Highsmith 1981), but each micro-crack is arrested by other constituents

123

(such as fibers) or other microstructure (such as neighbouring plies). Fibers also break in many places, even along their length, since the surrounding material can transfer stress back into them quickly in a well designed composite system.

These complex 'damage modes' make it possible to use very strong, light, brittle materials in material systems that are strong, durable, and remarkably damage tolerant. Indeed, composites are the materials of choice for applications in which these qualities are controlling features. However, these complex damage processes also require new understandings and representations if we are to correctly anticipate the strength, stiffness, and life of components made from such material systems.

Several distinctive aspects of this physical behaviour are critical to our efforts to construct a modelling approach:

1. Brittle composite systems fail by a statistical *accumulation* of defects.

2. Large changes in material properties such as stiffness, strength, and electrical thermal/conductivity occur in many such systems before rupture is eminent. 'Failure' may be due to any combination of these changes, and the variations in response during life is a critical part of the problem.

3. Many different failure modes are typically demonstrated by a material system, depending on the nature and history of the applied conditions.

4. Intrinsic material system strength is generally controlled by highly local conditions and processes, often at the fiber/interphase/matrix level.

An extensive (albeit, not well codified) body of observations that support these paradigms is available in the literature (ASTM 1982; Elsevier Science Publishers 1990; Razvan & Reifsnider 1991). For our purposes, we draw special attention to the phenomenological behaviour usually classified as creep, fatigue (static and cyclic), and ageing. We will then attempt to construct a representation of the combined effects of these classes of behaviour.

Creep is generally discussed as time-dependent deformation under a constant applied load. In composite systems, creep can be caused by creep of the constituents and by creep of the interface or interphase regions between the constituents (Chapman & Hall 1992). Many descriptions of this behaviour have been constructed (Dillard 1990; ASTM 1983). For our purposes, two aspects of this behaviour are important to our models, the stiffness changes involved, and the temperature dependence of the phenomenon. A typical stiffness change as a function of time for a polymer (epoxy) based system is shown in Figure 1. That figure actually shows a master curve that estimates the changes in the transverse modulus E_{22} and the shear stiffness, G_{12} as a function of log-time. If the material is linearly viscoelastic, the temperature dependence may be added by shifting the time axis as a function of temperature, based on laboratory characterizations (Morris et al. 1979). Hence, for our present discussion, we will account for creep effects with time-dependent variations of matrix-controlled stiffness as a function

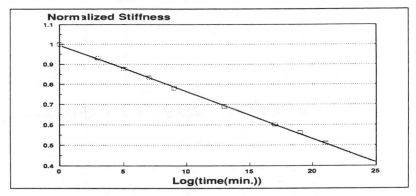

Figure 1. Creep-stiffness change in a polymeric composite.

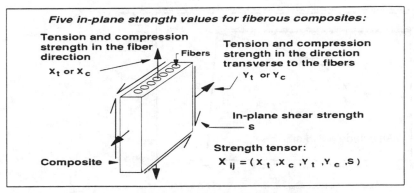

Figure 2. Planar 'principle strengths' for a fibrous composite.

of ambient temperature. In other settings, we have also included non-linear effects such as stress level dependence and physical or chemical ageing in these representations; more will be said of this below.

A second time-dependent behaviour is static fatigue or 'creep rupture', a time-dependent fracture under constant applied conditions. We will represent this behaviour by changes in material 'principle strengths'. In general, the strength of fibrous composite materials is represented by an array of values that reflect the anisotropic nature of the materials. (Fig. 2) For planar materials, the tensile strength and compressive strength in the fiber direction and transverse to the fibers, and the in-plane shear strength are required for a complete answer to the question of 'how strong is this material'. Out-of-plane values may also be needed. These values are typically combined with stress components in comparable directions in a 'failure criterion' (such as a Tsai-Hill, Von Mises, or Tsai-Wu criterion, determined by the operative damage and failure modes) to estimate component strength. A

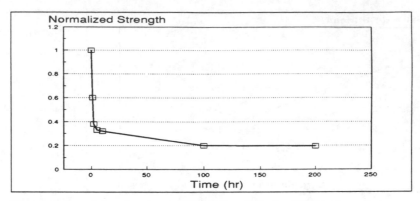

Figure 3. Strength reduction at 950°C for a Nicalon/SiC system.

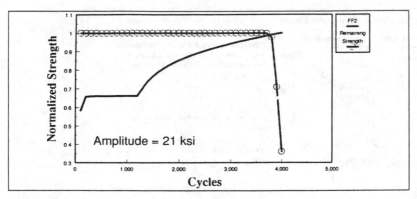

Figure 4. Creep compliance for polymer composite showing the effect of ageing for long time periods.

great many creep rupture representations (such as the Larson-Miller equations) are in the literature, but they only model the time to failure as a function of load and temperature. For our purposes, we will use those representations to scale strength loss for load level and temperature, but we will collect data to establish the degradation of the principle strengths as a function of time for representative cases of interest. These will be fundamental inputs to our model. Of course, such data are environment dependent. An example is shown in Figure 3.

We also address the question of cycle-dependent fatigue behaviour. As mentioned earlier, the effects of cyclic damage development are generally quite unusual in brittle composite systems. Widely distributed matrix cracking can cause stiffness changes of the order of 5-30%, depending on the material system (Highsmith & Reifsnider 1982). These stiffness reductions are not isotropic; different stiffness components change differently. In polymer matrix systems, for example, fiber-direction stiffness may change by very small amounts, while stiffness transverse to the fibers, and in shear, may change by large amounts. In ceramic matrix composites, stiffness change in the fiber direction may be substantial. Characterization of these changes (or accurate estimation from models) is essential. We will use these characterizations, directly, in our models.

Finally, we must consider ageing, in various forms. Polymer composites adjust the free volume as a function of time and temperature, and change their quasi-static and creep behaviour accordingly (Seitz 1993). An example of the effect of ageing on the creep compliance of a polymer composite is shown in Figure 4. In such a case, the effect of physical ageing can be modelled by an adjustment of the stiffness-time-temperature relationships discussed earlier. However, chemical ageing may also occur, wherein such things as oxidation may reduce stiffness or strength by a process that may be controlled by chemical reaction rates or by diffusion rates. Hence, the strength evolution relations discussed earlier may also be affected.

2 DAMAGE TOLERANCE AND DURABILITY CONCEPTS

Durability and damage tolerance are critical to the design of composite structures. Damage tolerance is the approach often required for the certification of safety-rated structures such as aircraft components; durability has been identified as one of the most important technical drivers for the design of major high performance composite structures such as high speed transportation devices (National Academy Press 1990, 1988). Of course, there are many nuances in the definitions of durability and damage tolerance. However, the fundamental ideas will be discussed as shown in Figure 5.

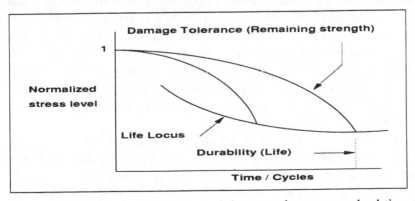

Figure 5. Definitions of durability and damage tolerance on a load-time (or life) diagram.

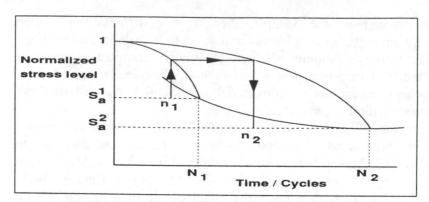

Figure 6. The concept of remaining strength as a damage metric.

Damage tolerance is the remaining strength after some period of service, and durability, in general, m has to do with how long the component will last, i.e. with the life of the structure. In this context, durability is often discussed in terms of the resistance or susceptibility to damage initiation. Both concepts imply that the subject component is being exposed to applied conditions such as mechanical loading and environments such as elevated temperature and chemical agents over long periods of time that are typical of the projected service life of the component.

We begin the formulation of our approach by selecting remaining strength as a damage metric. Figure 6 shows the essence of this argument. We assume that remaining strength can be determined as a function of loading level (with other applied features constant), or that it can be predicted by models such as the present one. For two such levels, a given fraction of life corresponds to a certain reduction in remaining strength. We claim that an 'equivalent' fraction of life at another load level would give the same strength reduction, and that the remaining life at the second load level is simply the number of cycles needed to further reduce the life from that point to the (second) level of applied cyclic loading. Hence, the effect of several increments of loading can be summed by adding their respective strength reductions. However, the strength reduction curves are, in general, non-linear, and highly dependent on all applied conditions, so that the resulting remaining strength and life calculations are highly non-linear, and path as well as history dependent.

The next step in our approach is to postulate that *normalized* remaining strength (our damage metric) is an internal state variable for damaged material systems, in the sense of Kachonov and others (Kachonov 1986). Of course, 'failure criteria' such as Hill or Von Mises equations, or more recent forms such as Tsai-Hill and Tsai-Wu criteria, are formed by scaler combinations of the ratios of stress in principle material directions divided by the principle material strengths in those directions, so we will use those failure functions to represent our normalized remaining strength, chosen on the basis of their suitability for the failure mode that controls the remaining strength and life for the component and conditions of interest for a given situation. Then, retiring to our state variable formulation we begin with the premises that:

1. Although fracture strength is a scaler, it is a function of the material strength tensor and stress tensor for the instant of definition (analogous to stiffness, conductivity, ... which are collectively defined by a state of material *in relation to* an applied (tensor) condition).

2. We construct a state variable with the familiar scaler 'continuity', ψ, defined as $(1 - Fa)$ (or more generally, some function of Fa) where Fa is a mechanistically or phenomenologically based failure criterion defined in a 'critical element' (which defines our domain) for which rupture defines rupture of the system.

With these definitions, we claim that:

1. (instantaneous tangent stiffness) = $E(\psi)$, i.e. the change in any stiffness component can be defined in terms of ψ (Kachonov 1986).

2. Remaining strength and life can be defined in terms of ψ (as we will show below).

Then using standard thermodynamics arguments, the Helmholtz free energy $f = f(\psi, \varepsilon_{ij})$, becomes $f = U\psi$ as in classical Kachanov (1986) theory. Hence:

$$df = \sigma_{ij}d\varepsilon_{ij} - Qd\psi \quad \text{where} \quad Q = -U \tag{1}$$

where U is the undamaged strain energy density and Q is associated with the increment of *entropy* created by damage (and has the nature of energy released by degradation of the material state).

The central issue is the kinetics equation. We assume that the kinetics are defined by a specific damage accumulation *process* for a specific fracture mode, and define rates for all such processes of interest. We start with the most general, common kinetic equation (a power law), such that:

$$\frac{\delta\psi}{\delta\tau} = A\psi^n$$

where τ is a normalized, generalized time variable (monotonic increasing), and n is a material constant.

Generally,

$$\tau = \frac{\tau}{\hat{\tau}}$$

where $\hat{\tau}$ is the characteristic time constant for the process at hand. $\hat{\tau}$ could be a creep time constant (Christensen 1981), a creep rupture life, a fatigue life, etc. such that, for example,

$$\tau = \frac{n}{N}$$

Then:

$$\int_{\psi^o}^{\psi^i} d\psi = A \int_0^{\tau^i} (\psi(\tau))^n \, d\tau \tag{2}$$

The left hand side is

$$\psi^i - \psi^o = 1 - Fa^i - 1 + Fa^o \tag{3}$$

If we set $A = -1$ (since continuity is decreasing), and approximate $n = 1$, then

$$\Delta Fa = Fa^i - Fa^o = \int_0^{\tau^i} (1 - Fa(\tau)) \, d\tau \tag{4}$$

which is the change in the instantaneous value of the failure function for this process.

Then we define a 'residual strength' Fr such that

$$Fr = 1 - \Delta Fa(\tau) \implies Fr = 1 - \int_0^{\tau} (1 - Fa(\tau)) \, d\tau \tag{5}$$

where all quantities are defined in the critical element and for the process characterized by the characteristic time $\hat{\tau}$. A degenerate special case of Equation (5) occurs for

$$\tau \to \frac{n}{N}; \quad \hat{\tau} = N$$

for which

$$Fa(\tau) \to \frac{S_a}{S_u}$$

the ratio of unidirectional applied stress over unidirectional strength, where upon

$$Fr = \frac{S_r}{S_u}$$

and Equation (5) integrates to

$$\frac{Sr}{Su} = 1 - \left(1 - \frac{Sa}{Su}\right)\frac{n}{N} \tag{6}$$

a linear degradation of strength from initial to final value. Equation (6) is also an identity in the sense that it satisfies the end points of the residual strength curve, i.e. it is correct at the limits. In general, however:

$$Fa = Fa\left(\frac{\sigma_{ij}(n)}{X_{ij}(n)}\right); \quad N = N(n) \tag{7}$$

or

$$Fa = Fa\left(\frac{\sigma_{ij}(t)}{X_{ij}(t)}\right); \quad N = N(t) \tag{8}$$

If we introduce explicit time dependence in our basic kinetic law to read

$$\frac{\delta\psi}{\delta\tau} = \psi j \tau^{j-1}$$

we obtain the final kinetic equation in the form

$$Fr = 1 - \int_o^\tau \{1 - Fa(\tau)\} j(\tau)^{j-1} d\tau \tag{9}$$

where j is a material constant. This is the general form we will use to calculate the remaining (global) strength.

Now our problem has been reduced to calculating inputs to Equation (9). However, we still need to interpret the quantities that appear in that equation in terms of the physical details we discussed earlier. For that purpose, we introduce the 'critical element' concept (Reifsnider 1991). As shown in Figure 7, we claim that since damage is distributed uniformly during the accumulation process, we can chose a representative volume (in the usual continuum sense) for our analysis. However, we make that selection based on the failure mode that we wish to analyze, i.e. we set a boundary value problem for each distinct failure mode, and we set that problem on the basis of the local material state and stress state immediately prior to the failure event, i.e. in the fully damaged condition. In that condition, (based on careful laboratory examinations,) we can identify part of the representative volume as 'critical', in the sense that fracture of that part induces fracture at the global level. The other material elements in the representative volume are said to be 'subcritical', in the sense that their fracture does not cause fracture of the global component

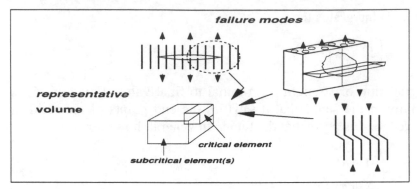

Figure 7. Representative volume, critical element, and subcritical elements as a basis for setting the mechanics boundary value problem.

All of our remaining strength calculations will be conducted in the critical element based on the boundary value problem set on the representative volume. For fiber dominated tensile fracture modes, for example, it has been found that global initiation of fracture is induced by a critical number of adjacent local fiber fractures, a small number of the order of 4-10 for many material systems (Tamuzs 1981). In that case, the representative volume would be a region large enough to contain such a group of broken fibers and enough surrounding material to recover 'undisturbed average' local stress fields and behaviour, and the critical elements are simply the fibers. Other examples are just as logically determined (Reifsnider 1992).

Now we can make a clear connection to the phenomenological behaviour we discussed earlier. The stiffness changes caused by creep and ageing, as we discussed, and other stiffness changes caused by micro-cracking during cyclic loading will act to change the tensor stiffness values used to calculate the local stress fields in the critical element. And if we can determine the rate of those different stiffness changes, we can calculate the correct local stresses as a function of cycles and time as inputs into Equation (9). In fact, we can calculate the stresses based on representations at various levels of the microstructure of the composite system. If we wish to use stiffness evolution information at the ply level, we can use laminated plate theory, or if we have information about the stiffness changes in the constituents (like the viscoelastic flow of the matrix, for example), we can use micromechanics to calculate the redistribution of local stress in a very direct way. This is, in fact, a major feature of this approach; the redistribution of stress is typically large, and plays a critical role in the proper determination of remaining strength and life.

Having discussed how stiffness changes alter the local stress state as a function of time and cycles, we turn to the question of changes in material state. That brings us to the question of the interpretation of the principle strengths that appear in the failure function in Equations (7) and (8). Another distinctive feature of the present approach is that we use micromechanics to represent the principle material strengths in terms of the properties, geometry, and arrangement of the constituents and the interphase regions between them. This approach has many advantages, one of which is a critical feature to the present approach. Many of the processes that control behaviour such as creep, creep rupture, and fatigue degradation are highly non-linear in their dependence on the applied conditions and history. A major question for any modelling attempt like the present one is 'how can one correctly combine the effects of these phenomena when they are happening at the same time?' The present approach claims that the answer to that question is, 'by letting the changes in the properties (geometry and arrangement, if appropriate) of the constituents and the interphase regions between them enter, directly, the micromechanical models of principle strengths and the calculations of local stress state as continuous functions of time (or cycles) in Equations (7) to (9), so that their collective effect is summed, incrementally, by the integration to the present time of evaluation. Hence, the physics and mechanics of the quantities defined in Equation (8) correctly combine the effects, without 'artificial' algorithms. However, it should be noted that great care must be taken to ensure that the rate equations that must be measured in the laboratory (and all of the constitutive behaviour, for that matter) under combined conditions are determined in such a way that the influence of independent external variables such as temperature and stress (or strain) are properly isolated.

To illustrate the use of micromechanics for the representation of principle strengths, a brief outline of the derivation of an abbreviated and simplified form of the tensile strength in the fiber direction for a fiber-dominated material will be presented below. This version of the model recovers many of the important features, but is a linear model of a process that is often highly non-linear. We have presented non-linear versions earlier (Aveston et al. 1971).

Figure 8 depicts the problem at hand. The stress in the broken fiber build back up to the undisturbed level by shear transfer from the surrounding matrix, composite, and interphase region. That rate of build-up is directly proportional to the stress concentration in the next nearest fibers; if the build-up occurs over a short distance, the stress concentration in the neighbouring fibers is great and they tend to break causing very brittle composite behaviour. However, if the build-up occurs over a large distance (i.e. very low fiber-matrix coupling by the interphase region) the strength of each fiber is lost completely when the first local fiber break occurs. Most fiber strength models consider only limited aspects of this problem. For ex-

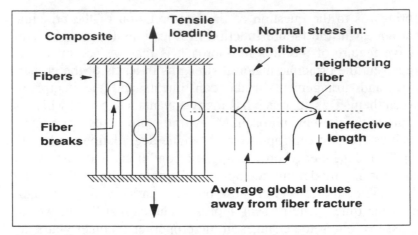

Figure 8. Tensile strength problem in the fiber direction.

ample, any 'complete' model would indicate that the fiber-matrix interaction should be compliant or weak enough to avoid transmitting a stress concentration to the next unbroken fiber, but stiff and strong enough to provide some shear transfer to the matrix so that the fibers can break more than once, i.e. so that the composite is stronger than a simple rope of fibers without any matrix. Surprisingly, very few models predict such an 'optimum' condition in terms of the material properties and local geometries of the problem (Gao & Reifsnider 1993). Batdorf (1982) has considered the effect of local stress concentration and the statistical distribution of strength of the fibers, but his model does not treat the effect of changing ineffective length at the local level as a function of load level associated with fiber-matrix sliding (Batdorf 1982). Curtin (1991) considers the latter aspect, but the question of stress concentration is not addressed (Curtin 1991). A model of tensile strength which builds on those three models is outlined below. A more complete derivation will be given in a subsequent paper.

We begin by writing the probability of failure of the fibers in a composite in terms of the applied stress, in the following form

$$P(\sigma) = 1 - \exp\left[-\delta k \sigma^m\right]$$

where δ is the local ineffective length normalized by the reference length, L, $k = 1/\sigma_o^m$ (where σ_o is the Weibull characteristic strength of the fibers), and m is the Weibull shape parameter of the fiber strength distribution. Then using the development of Batdorf, one can write

$$Q_i = Q_i n_i k \lambda_i (C_i \sigma)^m$$

in which Q_i are the number of i neighbouring broken fibers (or 'iplet'), n_i is the number of nearest neighbours around each iplet, i is a local stress redistribution measure, and C_i are the local stress concentrations caused by an iplet. Now we depart from the Batdorf (1982) approach and claim, first, that the condition of unstable fiber fracture (composite system fracture) can be assessed by requiring that *all* next nearest fibers break around any group of n broken fibers, i.e.

$$Q_{n+1} = Q_n n_n => Q_n n_n k \lambda_n (C_n \sigma_c)^m = Q_n n_n$$

when the applied stress has reached the critical level, σ_c. Then we can solve for that critical 'composite strength' to obtain

$$\sigma_c = \frac{1}{(k\lambda_n)^{1/m} C_n} = \frac{\sigma_o}{\lambda_n C_n}$$

Now if we substitute the earlier form for λ_n, and simplify the expression, we obtain

$$\sigma_c = \frac{\sigma_o}{\left[\left(\dfrac{\delta_n}{m+1} \right) (C_n^m + C_n^{m-1} + ... + 1) \right]^{1/m}}$$

We note in passing that if there are no local stress concentrations, the sum in the denominator of that expression becomes $m + 1$, and the strength expression becomes the classical bundle strength estimate for that limiting condition. We also note that this simple expression predicts a maximum (optimum) in the strength of the composite as the ineffective length and the local stress concentration are varied, as shown by Figure 9. Hence, such an expression correctly follows the physics that we know control the situation, and guides us in the choice of material parameters to avoid excessive local stress concentrations on the one hand and excessively long ineffective lengths on the other.

However, at this point we encounter the limits of current mechanics models in that no closed form solutions for the stress concentrations or the ineffective lengths are available for the case when several fibers are broken locally. For our case, we can approach the question of the ineffective length by using Curtin's (1991) model to write

$$\lambda_n \to \frac{2}{L} \left(\frac{\hat{D} F_f \varepsilon}{4\tau_o} \right)$$

(with the definition that $\hat{D}$ is an 'effective' fiber diameter that has been adjusted to account for the fact that multiple fibers are broken in the critical

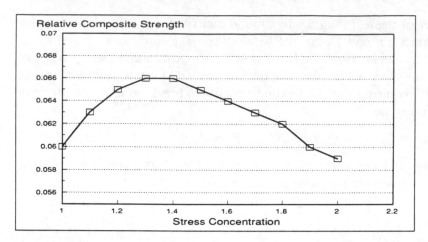

Figure 9. Global composite strength as a function of local stress concentration caused by *n* broken fibers.

region that initiates unstable fracture). Then if we claim that when $E_f\varepsilon \to \sigma_o$ the applied stress reached the critical level, σ_c we obtain the estimated strength expression

$$\sigma_c = \sigma_o^{\frac{(m-1)}{m}} \left(\frac{L2\tau_o}{\hat{D}}\right)^{1/m} \frac{(1+m)^{1/m}}{[C_n^m + C_n^{m-1} + \ldots + 1]^{1/m}}$$

that compares nearly term by term with the expression given by Curtin (1991), except for the dependence on the stress concentration, and the effect of fiber slipping which he incorporates. If we adopt those two 'correction factors' from Curtin's (1991) work, the final expression becomes

$$\sigma_c = \sigma_o^{\frac{m}{m+1}} \left(\frac{2\tau_o L}{\hat{D}}\right)^{\frac{1}{m+1}} \left(\frac{2}{m+2}\right)^{\frac{1}{m+1}}$$

$$\left(\frac{m+1}{m+2}\right) \frac{(1+m)^{1/m}}{[C_n^m + C_n^{m-1} + \ldots + 1]^{1/m}}$$

This expression has numerous advantages for the estimation of static strength, and for use in the estimation of remaining strength under long-term mechanical and thermal loading. Most important, it is explicit in the dependence on the micro-properties that control that strength or remaining strength, such as the interfacial shear stress/strength, τ_o, and the local stress concentration, C_n. In principle, the local stress concentration can be calculated, so that the only unknown for a given material system that is difficult to obtain is the critical number of broken fibers in a local region when unstable fracture ensues. As it happens, that number has been measured for some systems, and varies over a small range for those cases (Razvan 1991; Tamuzs 1981). Most important, our closed form strength expression above shows an optimum ('correct') choice for the material properties involved, including the interfacial strength that should be used to balance the local stress concentration against long ineffective lengths which reduce the bundle strength of the system. Other micromechanical models have been developed for other failure modes such as compression.

3 PERFORMANCE SIMULATION –THE MRLIFE CODE

Over the last twelve years or so, most elements of the approach described above have been assembled in a simulation code series, called MRLife. This copyrighted code is updated and released roughly once each year, and has been used to analyze many practical examples. The code has been (or is being) incorporated into the design codes of several major corporations, in various forms. We will use the MRLife simulation code to demonstrate the utility of the approach with examples and comparisons with data.

Figure 10 reviews the inputs to Equation (9), as discussed earlier. The code uses this equation to calculate remaining strength in a manner outlined in the flow chart in Figure 11. Laboratory observations are used to establish the failure mode for a given situation and material system of interest. The failure mode defines the critical element, and the constitutive equations that are needed to define the state of material in that critical element, as well as the rate equations that are needed to define the changes in the properties of the constituents as a function of applied conditions and history. An appro-

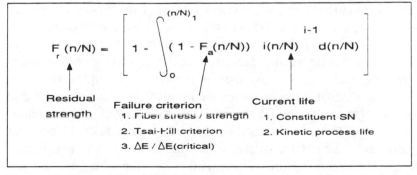

Figure 10. Interpretation of the inputs to the equation used to calculate the remaining strength of the composite representative volume.

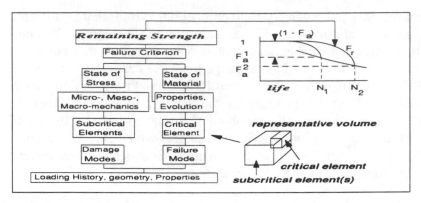

Figure 11. Conceptual flow diagram of the operation of the MRLife performance simulation code.

priate failure criterion is chosen, and the material states and stress states are entered into that criterion as a function of cycles and time to estimate remaining strength, using Equation 9.

To illustrate the utility of the method, two short examples will be given. A more complete presentation of the details of these examples will be presented in subsequent publications. Only an outline of the physical details can be presented in this summary of our model.

Nicalon/SiC ceramic matrix composites (CMC's) are primarily intended to be used as high temperature materials in applications such as heat exchangers and jet engine components. Most such materials now being used are designed to be fiber-dominated. When loaded, these composites exhibit matrix cracking at relatively low stress levels. Fracture toughness of the composite is achieved by creating a weak interface between the fiber and matrix so that matrix cracks do not fracture the fibers. The resulting damage state is highly distributed, as discussed earlier.

In the micro-kinetic model discussed above, matrix cracking and fiber fracture enter the calculation of local stress by causing stiffness changes and stress redistribution; usually, the stress in the critical element increases with loading history as a result of these events. Also, as oxygen enters the matrix cracks, oxidation of the fibers and depletion of the pyrolytic carbon interface (at high temperatures) causes loss in material (principle) strengths (Frety & Boussuge 1990). (The non-stoichiometric fibers undergo a rapid loss of static strength in this environment (Pysher et al. 1989)). Rate equations for this strength loss are obtained from creep-rupture tests. Hence, the characterization data needed are stiffness loss due to cyclic loading, reduction in life due to creep-rupture mechanisms, and the changes in fiber strength and stiffness with temperature. Stiffness change is a function of

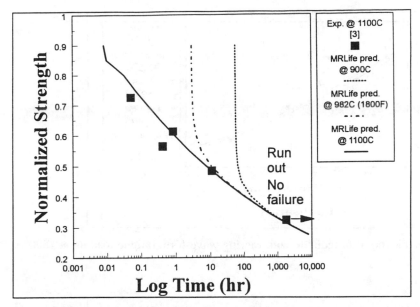

Figure 12. Creep-rupture data for enhanced Nicalon/SiC composites in air.

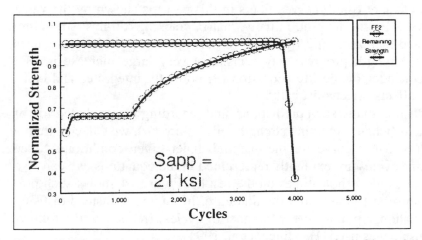

Figure 13. Predicted remaining strength curve (top) and value of local failure function (bottom) for enhanced Nicalon/SiC under tensile fatigue loading at 1800°F.

stress, temperature and cycles, and strength loss is a function of stress, time, and temperature.

Figure 12 shows typical creep rupture data, and also shows predicted curves from Equation (9) for three temperatures (including a temperature close to the case used to model the data.) Our kinetic approach is used to estimate the results at other temperatures.

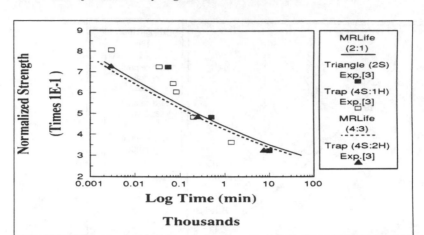

Figure 14. Prediction of fatigue life with varying wave-form fatigue loading at 1800°F, in air.

When stiffness change is added, Equation (9) predicts the results shown in Figure 13. The local change in the state of stress and state of material in the critical element (the 0 degree fibers in this case) is shown by the value of *Fa*, called 'FF2' in the plot. Initially, matrix cracking 'dumps' stress into the fibers and increases FF2 by about 10 percent. Then oxidation reduces the material strength progressively, causing a very large increase in FF2 over the remainder of the life. Equation (9) correctly integrates and combines these effects, mechanistically.

By conducting many such predictions, and recording the predicted life of the specimens (when remaining strength falls below *Fa*), we can construct a stress-life (or '*S-N*') curve for the material under given conditions. Since both time and cycles are explicitly represented in the equations, we can also calculate the effect of hold-times on the results. Figure 14 shows such predictions compared to data from the literature. In this case, instead of using the usual relationship that time = (number of cycles)/frequency, the following relationship was used (Headinger et al. 1994).

$$\text{Stress} * \text{Time} = \text{Cycles} \int \text{Stress} \, (t) \, dt$$

Fatigue simulations were made for triangular and trapezoidal loadings for which the integral of stress (*t*) were 1/2 and 3/4, respectively. Agreement is seen to be very good for low stress levels, and not as good for very high stress levels. Hold times do change the predictions and the data, as expected.

A second example is presented, using data from several sources for comparison to predictions of remaining strength and life of a carbon/PEEK

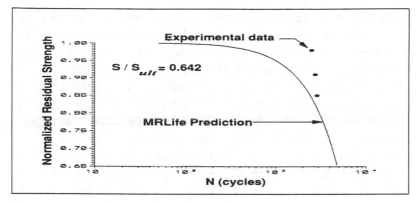

Figure 15. Comparison of predicted and observed residual strength for a stress amplitude of 0.7 of the ultimate strength.

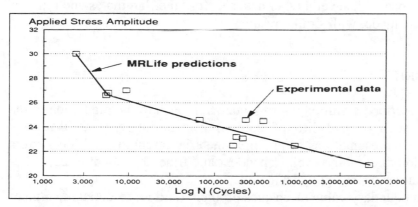

Figure 16. Comparison of predicted and observed fatigue lives of center notched quasi-isotropic coupons under fully reversed fatigue loading.

system, specifically, an Aromatic Polymer Composite (APC-2) with a Poly Ether Ether Ketone (PEEK) matrix reinforced with AS-4 carbon fibers. Unidirectional fatigue data were taken from Picasso & Priolo (1988). In order to use the form of the fatigue rate equation mentioned earlier, two regions of behaviour were identified, and the following constants were used in those regions.

$$A_n = 1.3126; \quad B_n = -0.1818; \quad P_n = 1.0$$
$$A_n = 0.7865; \quad B_n = -0.0425; \quad P_n = 1.0$$

On the basis of these data alone, and rate equations for the rate of micro-crack development in the neighbourhood of a center hole obtained from earlier measurements on a carbon-PMR-15 system, predictions of the remaining strength and life of laminates having a quasi-isotropic stacking sequence were made, for the case of fully reversed fatigue loading ($R = -1.0$). The predictions were made for room temperature conditions, so that creep or creep-rupture effects were not active. Hence, the code calculates the effect of the off-axis plies, applies a Whitney/Nusimer-type failure criterion to a 'characteristic' region (with dimensions of 4.45 mm), and estimates the remaining strength. Many such predictions have been compared to experimental data. An example of the agreement obtained is shown in Figure 15.

Each of the data points shown represents only one test, but many other comparisons were made for other cases. The code predicted failure in compression, which was observed in the laboratory. An array of these predictions was made and used to construct a predicted fatigue life (*S-N*) curve, and compared to the test data of Simonds & Stinchcomb (1989). The results are shown in Figure 16.

Comparisons with several other data sets for other laminates and types of loading were made, with similar results.

4 CLOSURE

We have described a unified micro-kinetic approach to the prediction of the remaining strength and life of composite material systems that uses mechanistic representations of principle material strengths and local stress redistribution to combine cycle dependent and time dependent effects. Distinctive features of the approach include:

– Sets a boundary value mechanics problem (and representative volume) on the state of stress and state of material defined by *incipient fracture* for each failure mode;

– Identifies a representative volume element, critical element, and sub-critical elements for each distinct failure mode;

– Defines the damage-related state variable in Gibbs free energy of the system as a failure function written in terms of ratios of local stress components to corresponding principle material strength components;

– Defines the damage metric in terms of remaining strength;

– Uses mechanistic micromechanics representations for principle material strengths to relate micro-details of properties, performance, geometry, and arrangement of constituents and interphase regions to global performance;

– Predicts remaining strength and life in terms of the properties, performance, geometry, and arrangement of the constituents and the interphase regions between them.

Examples were presented for polymer and ceramic based composites. Comparisons with data (including earlier examples) suggest that this approach offers a viable method of predicting the remaining strength and life of composite systems subjected to combinations of mechanical, thermal, and chemically aggressive environments. The approach is a framework that incorporates and combines representations of physical behaviour. The predictions of the model will improve as those representations (and the understandings that support them) improve. There is special need for advances in the understanding and modelling of time dependent behaviour including creep, creep rupture, and ageing, especially at the basic level. There is also a need to better understand how the point-wise predictions of this model can be used to estimate the reliability and damage tolerance of an engineering structure. Both of these problem areas (and others) are being addressed in our laboratory, but they would benefit from broader attention by the general community.

REFERENCES

Aveston, J., Cooper, G.A. & Kelly, A. 1971. Single and multiple fracture. *Proc. Cont. on the Properties of Fiber Composites.* IPC Science and Technology Press, Guildford.

Batdorf, S.B. 1982. Tensile strength of unidirectionally reinforced composites. *J. of Reinforced Plastics and Composites.* Vol. 1.

Ceramic Matrix Composites: 158. 1992. R. Warren, Ed., Chapman and Hall. New, York.

Christensen, R.M. 1981. Lifetime predictions for polymers and composites under constant loads. *J. of Rheology* 25, no. 5: 517-52

Curtin, W.A. 1991. Theory of mechanical properties of ceramic-matrix composites. *J. Am. Ceram. Soc.* 74, no. 11: 2837-2845.

Damage in Composite Materials: 103-117. 1982. ASTM STP 775, American Society for Testing and Materials, Philadelphia.

Dillard, D.A. 1990. Viscoelastic behavior of laminated composite materials. In *Fatigue of Composite Materials:* 339-389. K.L. Reifsnider, Ed., Elsevier Science Pub., New York.

Fatigue of Composite Materials 1990. K. L. Reifsnider, Ed., Elsevier Science Publishers, Amsterdam.

Frety, N. & Boussuge, M. 1990. Relationship between high-temperature development of fiber-matrix interfaces and the mechanical behavior of SiC/SiC composites. *Comp. Sic. & Tech.* 37: 177-189.

Gao, Z. & Reifsnider, K. 1993. Micromechanics of tensile strength in composite systems. In Fourth Volume, *ASTM STP* 1156: 453-470. W.W. Stinchcomb & N. F. Ashbaugh, Eds, Am. Soc. for Testing and Materials, Philadelphia.

Headinger, M.H., Roach, D.H. & Landini, D.J. 1994. High-temperature fatigue of ceramic matrix composites. Presented at AeroMat '94, Los Angeles (private communication).

Highsmith, A.L. & Reifsnider, K.L. 1982. Stiffness-reduction mechanisms in composite laminates. In *Damage in Composite Materials:* 103-117. ASTM STP 775,

American Society for Testing and Materials, Philadelphia.

High-Temperature Materials for Advanced Technological Applications 1988. NMAB-450, National Academy Press, Washington, D.C.

Kachonov, L.M. 1986. *Introduction to Continuum Damage Mechanics,* Martinus Nijhoff Pub., Dordrecht, Netherlands.

Life Prediction Methodologies for Composite Materials 1990. NMAB-460, National Academy Press, Washington, D.C.

Long-Term Behavior of Composites 1983. ASTM STP 813, American Society for Testing and Materials, Philadelphia.

Morris, D.H., Yeow, Y.T. & Brinson, H.F. 1979. The viscoelastic behavior of the principle compliance matrix of a unidirectional graphite/epoxy composite. Virginia Polytechnic Institute and State University, VPI-E-79-9, Blacksburg, VA.

Picasso, B. & Priolo, P. 1988. Damage assessment and life prediction for graphite-PEEK quasi-isotropic composites. In *Pressure Vessels and Piping Division Publication PVP* Vol 46: 183-188. ASME, New York.

Pysher, D.J., Goretta, K.C., Hodder, R.S. & Tressler, R.E. 1989. Strengths of ceramic fibers at elevated temperatures. *J. Am. Ceramic Soc.* 72, no. 2.

Razvan, A. & Reifsnider, K.I. 1991. Fiber fracture and strength degradation in unidirectional graphite/epoxy composite materials. *Theoretical and Applied Fracture Mechanics* 16: 81-89.

Reifsnider, K.L. 1991. Performance simulation of polymer-based composite systems. In *Durability of Polymer-Based Composite Systems for Structural Applications:* 3-26. A.H. Cardon & G. Verchery, Eds, Elsevier Applied Science, New York.

Reifsnider, K.L. 1992. Use of mechanistic life prediction methods for the design of damage tolerant composite material systems. In *ASTM STP* 1157: 205-223. M.R. Mitchell et.al., Eds, American Society for Testing and Materials, Philadelphia, PA.

Reifsnider, K.L. & Highsmith, A.L. 1981. Characteristic damage states: a new approach to representing fatigue damage in composite laminates. In *Materials: Experimentation and Design in Fatigue:* 246-260. Westbury House, Guildford, Surrey, UK.

Seitz, J.T. 1993. The estimation of mechanical properties of polymers from molecular structure. *J. Applied Polymer Science* 49: 1331-1351.

Simonds, R.A. & Stinchcomb, W.W. 1989. Response of notched AS-4/PEEK laminates to tension/compression loading. In *Advances in Thermoplastic Matrix Composite Materials, ASTM STP* 1044: 133-145. G.M. Newaz, Ed., Am. Soc. for Testing and Materials, Philadelphia.

Tamuzs, V. 1981. *Proc. 2nd USA-USSR Symp. on Fracture of Composite Materials, Lehigh University.*

Importance of design aspects for safe structural integrity of composite structures

C. Hiel

NASA, Ames Research Center, Moffett Field, CA, USA and VUB, Brussels,Belgium

'A fundamental activity of engineering and science is making promises in the form of designs and theories which the users should approach with sufficient caution and healthy skepticism, for the history of science and engineering is littered with failed promises.'

Henry Petroski [1]

ABSTRACT: These notes have been compiled to provide the participants of the 5th. Special Chair AIB-Vinçotte 1995 with the author's perspective on the evolution of knowledge in structural integrity and on the importance of the design aspects when dealing with composite structures.

Four specific case studies, with which the author was personally associated, will be offered as a basis for clarification and discussion of the key issues.

The first case study concerns the sudden loss of structural integrity in the housing of a 17.5 KV circuit breaker. We will address the sudden loss of structural integrity which was caused by brittle failure. In passing general comments will be made on the design and durability concerns related to composite systems used in the electrotechnical industry.

The second and third example document hardware failures for a tilt-rotor aircraft and for the flex-beam of an X-wing aircraft respectively. Both applications are representative for today's advanced composites hardware. From a review of the failure modes conclusions of a general nature will be drawn as what to look for when inspecting composite systems. The attendees attention will be directed towards critical design steps which, when left out, can lead to very serious operational problems.

The last case study illustrates the design of a composite compressor blade. The attention will be drawn towards a durable joint design.

1 EVOLUTION OF KNOWLEDGE IN STRUCTURAL INTEGRITY

These notes have been compiled to provide the participants of the 5th. Special Chair AIB-Vinçotte 1995 with the author's perspective on the evolution of knowledge in structural integrity. A substantial part of this section is dedicated to the evolution of knowledge, in structural integrity of metals and a worked example has been provided in Appendix 1 to communicate

145

the essentials. Design aspects which were introduced for safe structural integrity, and which are used in today's structures, will also be mentioned in passing.

Once this modern approach to structural integrity for metal structures has been established, we will conclude with a discussion on key composite issues.

The Apurimac-Chaca suspension bridge [36], first erected in 1400, is commonly seen as one of the greatest engineering feats of the Inca society. A thousand Indians were needed to cut the agave to get the fibers to be spun into thick cables and to undertake the adventurous work of slinging the suspension bridge across the gorge. Although the fiber cables had to be changed every two years, the bridge survived from 1400 until 1880. This example indicates that the Inca's had definitely learned to live with the failure of a fibrous structure. This cable 'retrofitting' project, that was sustained over a period of 480 years, may very well be the first structural integrity program ever devised for the tension member of a fibrous load bearing structure. As will be mentioned in Section 2, several modern suspension bridges have also been in need of new cables after only a few years of service.

Structural integrity became a real problem with the invention of the steam engine and the locomotives it powered. By 1850 serious fatigue failures of locomotive axles were first experienced.

Prior to World War II, several welded Vierendeel-truss bridges in Europe failed shortly after being put into service. All the bridges were lightly loaded, the temperatures were low, the failures were sudden and the fractures were brittle.

Similar problems occurred when some of the famous liberty ships, which helped 'turn the tide' in World War II, developed fractures which started at square hatch corners or square cutouts. Design changes, involving rounding and strengthening of the hatch corners, removing square cutouts and adding riveted crack arresters in various locations, led to immediate reductions in the incidence of failures.

Out of these problems rose a systems approach to structural integrity which evolved for managing the interdisciplinary technologies that were required. Reference [2] contains a detailed account of key developments in this area i.e. fundamental understanding of materials by using X-rays, A.A. Griffith's classical 1923 publication on fracture, P.C. Paris's seminal paper on fatigue crack growth that was promptly rejected by three leading journals in the 1960's, because their reviewers uniformly felt that one could not correlate crack growth rates in the way he proposed etc.

In the mid 1950's a British made Comet aircraft had failed catastrophically, after only 3.682 hours of flying time [3]. An exhaustive investigation indicated that the failures initiated from very small fatigue cracks originating from rivet holes near openings in the fuselage. This caused explosive decompression at cruising altitude (10.000 m). This occurred, notwhitstanding the fact that engineers had designed the aircraft fuselage to withstand 2.5 times the cabin pressure demanded by the regulations put out by the air registration board [4]. New design aspects for safe structural integrity were formulated as a result to these failures. The aluminum that was specified for subsequent aircraft (i.e. the Boeing 707) was four and one half times as thick as the Comet's, to resist tearing. Moreover, at frequent intervals they welded to the inside of the skin 'tear stoppers' made of titanium. Thus, wherever a rupture might start, it would soon encounter a titanium strip blocking its deadly path.

In December 1966 the US Air Force experienced the loss on a F-111 aircraft that had only 100 hours of flight thereby killing both crew members. The crash was caused by an existing and undetected crack in the wing-pivot fitting which on the day of the crash resulted in loss of the wing during flight [5]. This small crack had been caused when the wing-pivot fitting was machined and drilled. It is unlikely that a crack of this nature would have spread in a softer material like aluminum, but the high strength steel used in the F-111 wing pivot happened to be very sensitive to small defects [6]. This problem led to the development of MIL-STD-1530 (Fig. 1a) and the USAF structural integrity program.

By the 1970's C.F. Tiffany of USAF [7] described a system of aircraft design and verification for structural integrity and durability which accepted fracture mechanics as a mature technology. It was set-up to prolong the life of weapon systems out of production and in active first line combat inventory. Its aim was also to improve performance and to include structural integrity know-how in future weapon systems. Figures 1a and 1b are from Tiffany's outline of command documentation and the interrelationship of data acquired from technical disciplines involved.

Appendix 1 contains a worked example, which applies Tiffnay's approach to demonstrate how the remaining life of a hypothetical metallic pressure vessel can be calculated. When working through this example one realizes that the modern approach to structural integrity does not even require the ability to predict the initiation of cracking. It is assumed that cracks exist. It is important to note that the loading on this hypothetical pressure vessel is very simple. Most engineering structures are subjected to a much more complex loading spectrum, such as the one shown schematically in Figure 1b, which continues to challenge currently existing life perdition methodologies.

Personal communication with engineers with as much as 50 years of design experience with experimental aircraft have convinced the author that

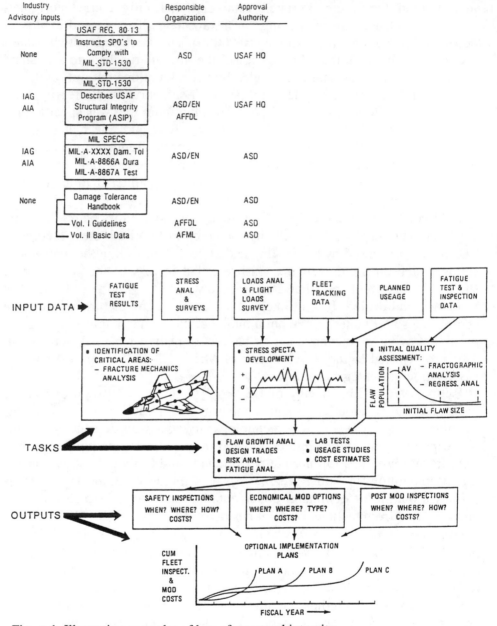

Figure 1. Illustrative examples of loss of structural integrity.

even life prediction methods for metals have not been adequately developed.

A world class helicopter designer has indicated to me that 'we cannot stay with metals as they currently are!' He bases his assessment on 40 years of keeping an eye on metal parts from the drafting board, through the workshop, through the test lab and through the flight tests. In all cases they followed the best practices they knew i.e. good surface finish, shot peening, nitriding, etc., to improve fatigue resistance; no cut threads, only rolled; inspecting metal billets for inclusions etc. The results were still disappointing as is made clear by the next two examples:

Retention straps in the form of stainless steel leaf bundles on the McDonnell XV-1 had fatigue allowables of $\pm$ 69 MPa. Net section nominal stress through the lug about average for steel lugs. However scaling the parts up to 2 $\times$ size for the XHCH-1, the allowable went down to around 38 MPa. These parts were 17-7 PH stainless steel with edges polished. Ultimate tensile strength (UTS) was around 1240 MPa: this is 38/1240 = 3% of the static strength-not too stunning a performance. But they have reported worse.

A cast-aluminum primary pylon structure on the XV-15 (see Section 4.3) cracked in seven places at nominal stresses around 3.5 MPa or 1.5% of ultimate. That's a Kf (stress concentration factor) of about 30. Admittedly, it was a poor design to use cast-aluminum. The manufacturer went further and used cast magnesium in adjacent parts. The management insisted that with the stresses so low, there was no need for fatigue testing, but engineers insisted on doing the test an many cracks occurred.

With 20/20 hindsight we can state that the evolution of knowledge in structural integrity has been spurred by a considerable number of catastrophic failures. Many of the interdisciplinary technologies required for the formulation of safe structural integrity programs were developed as a result of these failures. To date the problem of life prediction for metals has not been solved, although one has attacked it for the last 150 years. Some of the current problems were dramatically demonstrated at the Army's Helicopter Fatigue Methodology Specialist Meeting [37] in March 1980, in which various manufacturers, from around the world, presented vastly different conclusions to a simple type of 'canned', text-book-style sample problem. Details regarding this problem and its results will be briefly summarized during the presentation.

As further mentioned in Section 2, even today cracking continues to occur in the most advanced structural systems.

The structural integrity prediction given in Appendix 1 works well because it hinges on the concept of the length of a dominant crack. In composites we do not have such a simple and unambiguous manifestation of damage. The physics of crack growth, in a composite material, generally involves an interaction of a multiplicity of cracking modes i.e. matrix cracking, fiber matrix decohesion, fiber fracture, ply cracking, delamination etc. One therefore conveniently talks in terms of a damage parameter, rather than the length of a dominant crack. Reifsnider and his colleagues have taken the lead in attacking the core of this extremely complex life prediction problem by using a combination of deep physical insight and a rigorous mechanics formulation. Their results have been reported extensively throughout the literature but will not be listed, since Prof. Ken Reifsnider is personally addressing the problem during this lecture series [8].

It was demonstrated in Appendix 1 that proof testing of metal structures (i.e. a pressure vessel) is relied on to develop engineering confidence that the structure is unlikely to fail within a specified time and under a specified usage. Proof-testing has also become part of a number of structural integrity programs which were applied to composite structures. The US Air Force [9] has developed and applied a methodology consisting so-called 'wear-out testing' in combination with proof testing for the certification of outer composite wing of the A-7 aircraft, the composite empennage of the F-16, and the B1-A's horizontal tail.

The structural integrity programs for composites hardware have evolved into umbrella efforts, consisting of the following four components:

1. Prediction of operational life expectancy;
2. Diagnosis of failures;
3. Formulation of corrective action;
4. Importance (and improvement) of structural design aspects.

This list, which represents a considerable body of knowledge, incorporates virtually all the subject topics on which lectures were delivered as part of this 5th Special Chair AIB-Vinçotte 1995.

In this last lecture of the series we want to discuss practical examples related to element No 4 (importance of structural design aspects) It would however be irresponsible to isolate this topic from the other three, especially the lessons learned from failure are of major importance.

Note: Fracture mechanics techniques which rely on the toughness of metal alloys cannot presently be applied to helicopter dynamic systems because in most cases the resulting structure would be too heavy to get off the ground with the payload required to meet the economic break-even point. Oscillating strain levels would have to be reduced on the order of 50% in order for the crack propagation time to approach that needed to provide a realistic inspection period, or in some cases to return from a mission

2 THE COST OF FRACTURE

The Comet airliner, which was discussed above, represented England's technical and commercial capabilities as it was raising from the ashes of war. The fatal flaw which was hidden in the fuselage of the airplanes, and caused them to crash, discouraged the British from building further jets and instead they started to concentrate on turboprops. They would go on to build a variety of successful airliners using both turboprops and small jets.

Nevertheless they paid a heavy price for their slowdown in innovation, since they missed-out entirely on the brilliant economic future which belonged to the bigger jets [10].

In recent studies of the economic effects of fracture in the US [11] and Europe [12] the total loss to the gross economic product of advanced nations has been estimated to be 4%. For the US this is reported to be $120 billion [13] a large fraction of which (about $40 billion) can be avoided by applying the type of current-day understanding which we outlined in Appendix 1 and by implementing structural integrity programs.

Notwithstanding all this apparent knowledge one wonders about the ability to apply design aspects and lessons learned (aspect #4 which was given as part of the umbrella effort called SIP Structural Integrity Program) when moving from one structural design to the next.

The media continues to report problems. The August 1991 issue of Time [14] states the following; 'Air Force Officials have confirmed the discovery of tiny fissures in the wing carry-through bulkheads on a number of its 1.895 F-16's ($13.7 million a piece). Many of them will have to be modified or prematurely mothballed. They also found cracks in 37 of their 97 high tech B-1 bombers adding to the troubles of the controversial and expensive ($300 million a piece) craft. Meanwhile brittleness and cracks were found in the Navy's SSN-21 Seawolf class submarine. There are serious doubts whether the $2.5 billion sub-killing craft will ever go to sea.'

Although the examples mentioned cast doubt on the ability to produce a trouble-free structure, the structural integrity is undoubtedly present. Failures are diagnosed (element #2 in SIP) and corrective actions are being proposed (element #3 in SIP).

As an example the KC-135 tanker aircraft was first delivered to the Air Force in 1955 with an initial service life of 10.000 flying hours. Through modifications, including a reskinning effort the KC-135's structural integrity is now assured to 36.000 hours. The B-52 was designed in 1948, and first flew in 1952. Two models are still in active inventory, although the original life was only 5.000 flying hours. The planes have undergone extensive structural modifications, including re-winging, structural fatigue modifications and a tail assembly modification [6].

These examples serve to show that prolonged structural integrity, even beyond the original design goals, is feasible if one is willing to pay for it. (the Inca's, in the example of the suspension bridge given in Section 1, paid for it during a 480 year period)

An interesting side aspect is that corrective action often involves the application of a crack arrestor which is a composite material. Reference [15] reports the use of a boron-epoxy wing splice which reduced stresses by 12% and multiplied fatigue life by 2, 4. Additional efforts have been reported in [16].

Despite maintenance programs, some of the worlds cable stayed bridges are in danger, because of fatigue. The cables of Venezuela's Maracaibo bridge built in 1962 have already been replaced once and are due for a second refitting. Likewise Germany's Kohlbrand bridge completed in 1975 needed a new set of cables after only three years; the problem was stress fatigue. The degradation of the cables spurred a search for better protection methods. Another persistent problem is corrosion. Engineers are experimenting with carbon compounds instead of steel. It is design, construction and maintenance that, in the end, combine to make a bridge work [35].

If there is one thing to be learned from the staggering cost of failure in metallic structures it should be this: 'Current best practices are not being used as much as they should because the Structural Integrity Procedures and Methodologies are considered the domain of the 'expert.' To make progress they should be placed in the hands of the people who design and build structures and components.'

Besides the cost of fracture one also needs to consider the cost of corrosion. The nationwide cost of checking and controlling corrosion in the US has been reported at $80 billion. Airlines alone have a $100 million annual corrosion problem [17].

On steel bridges, painting is a never ending chore. Scotland's Forth bridge e.g. has a full-time team of painters who apply 17 tons of paint per year in an ongoing maintenance program. A specific example that illustrates an advantage of composites will be given in Section 3.

Section 3 has been added in order to dispel the myth that commercial composite materials structures are expensive.

3 THE COST OF ADVANCED AND COMMERCIAL COMPOSITE MATERIALS AND STRUCTURES

So called 'advanced' composite structures may be expensive as is illustrated for the inlets-6 satellite, which contains a unique carbon/epoxy structure, with a total length of 12 m and a weight of nearly 1800 kg without fuel, costs roughly $77.000/kg or five times as much as gold. When it doesn't unfold in orbit, it becomes a $140 million flying junkyard.

Reference [18] gives the example of an AV8B aircraft that flew through a severe hailstorm at 600 knots. The aircraft had lost its canopy, metal inlets and had its composites teared. Nevertheless the plane was repaired due to the material's base repairability and toughness. An aluminum airplane flying through the same storm would have been scrapped.

Reference [19] reports that a rudder for a large commercial airliner has 17.000 elements when made with aluminum alloy. These are fabricated into 600 parts which in turn form a single rudder. On the other hand in the case of advanced composites only 4.800 elements are needed which are then assembled into 335 parts. The reduction of elements or parts by one-piece fabrication results in increased reliability and low production costs.

Extensive engineering experience has been gained in the use of commercial composites. Today, primary load bearing structures can be designed to be competitive in cost with equivalent structures made of steel or reinforced concrete. The example of high pressure piping is used to illustrate this fact:

The maximum allowable design strength of steel used to make pipe that meets the ASME B-31 pipe code is 138 MPa. This is only one third of the allowable design strength of so called Commercial Composite Materials (glass strand reinforced polyester or vinylester resin), which have an Hydrostatic Design Basis Strength [HDBS] of 414 MPa as determined from ASTM D2992-A cyclic test results. The density of steel is approximately four times that of Commercial Composite Material (CCM). Thus one kilogram of composites equals the strength of twelve kilograms of API grade 52 steel. Piping from CCM costs approximately $4.31/kg to manufacture. Steel made into pipe costs $0.88/kg

Of course cost vs. strength comparisons reflect only part of the total picture. Reliability as an end use structure also depends on the ease and reliability of joining and sealing the structural materials.

If one starts to consider more complicated parts it becomes important to realize that only 15% of the cost of a typical steel part is for steel. The other 85%, which is largely eliminated by using composites, is to achieve shape and finish. Additional life-cycle costs need to be factored-in to keep corrosion in check.

A turnkey factory for the fabrication of tanks can be set-up for $100.000. The Environmental Protection Agency (EPA) mandated leaking-tank replacement market is pegged at 900.000 installations over the next 3 years. The storage tank market (metal and composite) in the US is estimated at $10 billion/year. Current capacity in the US is only 30.000 tanks/year. It takes just 4 hours to yield a container of 11 m in length and 2.5 m diameter, weighing 2360 kg with a capacity of 55.000 liters. Material and labor costs come at $5,800 per unit. The cost per tank is thus $2.45/kg. The November 18, 1991 issue of Plastics News reported that a 22.700 liter fuel storage tank which had been buried for 26 years was dug up and tested. There was no leakage, no corrosion and therefore the tank was buried at a new site. Evidently double-walled fiberglass tanks are gaining favor.

The trouble with traditionally built houses is that they use a good chunk of their strength to keep themselves up. A typical house with 185 square meters of surface area in the US weighs 150 tons [20], which is 82 kg per square meter, and it sells for $135,000 in the West. The cost is therefore $0.90/kg.

A comparative house that was engineered and factory-build by Buck-minster Fuller [21] weighs 50 times less (3 tons). The materials used in such 'lightweight' house could therefore be allowed to cost $45/kg. The earlier example on fiberglass tanks indicated that they can be build for $2.45/kg which is more than 18 times less.

Pioneering work along these lines has been undertaken for the United Nations Industrial Development Organization (UNIDO) by the Belgian Professor Patfoort [22].

It should also be noted that carbon producers [19] have set their sight on producing a carbon fiber which costs less than $10/kg. They call this the ten dollar fiber which is being developed to open the automotive sector for advanced composites.

The recent earthquakes in Los Angeles and in Japan have cast doubt on the integrity of steel framework buildings [23]. This has caused the developers of 'Commercial Spaceport USA,' located at Vandenberg Air Force Base which is located on the California coast, to seriously consider building a sixteen story composite building to check-out rockets. The cost of this building has been estimated at $20 million, including the pultrusion machinery required for manufacturing. The cost for refurbishment, due to continuous corrosion in the coastal pacific environment, of an existing steel frame building has been estimated at $5 million. Rust is also known to cause a terrific explosion when it comes in contact with the rocket's hydrazine fuel. The composite building, including its joints, has already been designed and analyzed and the decision to go ahead and build it is expected to accelerate the market acceptance for composites in civil engineering applications by 30 years.

Research proponents and contractors have enthusiastically emphasized the most desirable features of composite materials usage, such as reduction in part count, reduced weight and reduced cost. Obvious problems such as the liability to split like wood in the weak direction, and to shatter under impact have sometimes been de-emphasized. Serious errors have also been committed repeatedly by designers analyzing the strong direction of the part and tacitly assuming no stress will ever occur in the weak direction.

The following four case-studies will point to typical weaknesses in design which cannot be tolerated if one needs to assure safe structural integrity of composite structures.

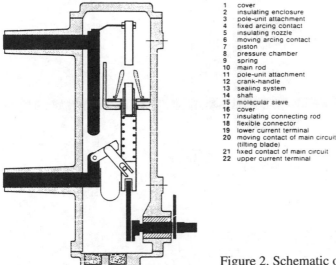

1   cover
2   insulating enclosure
3   pole-unit attachment
4   fixed arcing contact
5   insulating nozzle
6   moving arcing contact
7   piston
8   pressure chamber
9   spring
10 main rod
11 pole-unit attachment
12 crank-handle
13 sealing system
14 shaft
15 molecular sieve
16 cover
17 insulating connecting rod
18 flexible connector
19 lower current terminal
20 moving contact of main circuit
    (tilting blade)
21 fixed contact of main circuit
22 upper current terminal

Figure 2. Schematic of 17.5 KV circuit breaker.

4 FOUR CASE STUDIES

4.1 *Sudden fracture in the insulating enclosure of a 17.5 KV circuit breaker*

The unit which is schematically shown in Figure 2 consists of:

– A *main circuit* including the fixed contact and the self-wiping self-compensated blades making up the moving contact;

– A *breaking circuit* including the fixed arcing contact, moving arcing contact, main rod, flexible connector, moving piston and insulating nozzle directing the flow of Sulfur hexafluoride (SF6) gas towards the arc. The main circuit, designed for the continuous flow of the current, is distinct from the breaking circuit subjected to the arc;

– A *transmission mechanism conveying the mechanical energy* to the moving contacts, including the crank-handle, shaft, insulating connecting rod, main rod and spring;

– A *sealing system* highly reliable for a number of operations;

– An *insulating enclosure* made from a casting resin combined with a fiberglass weave, which encloses all active parts.

The operation of the unit is as follows:

– Figure 3a, the main contacts and the arcing are initially closed;

– Figure 3b, pre compression: As from the beginning of the movement the SF6 gas is compressed by the piston. The main contacts separate and the current flows via the arcing contacts which are still closed;

– Figure 3c, the arcing time: The arc then forms between the arcing con-

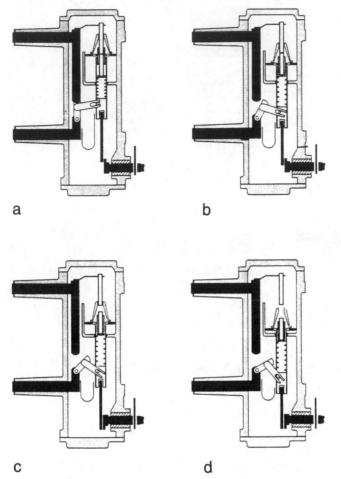

Figure 3. Schematic of operation of the unit.

tacts and the piston continues its downward movement. A small quantity of gas leaves the pressure chamber. It is directed by the insulating nozzle and injected on to the arc. The cooling of the arc is thus achieved through forced convection for the interruption of low currents, however during the interruption of high currents, it is thermal expansion which is responsible for the transfer of the hot gasses toward the cold parts of the pole-unit. Towards current zero, the dielectric strength between the contacts is recovered due to the intrinsic qualities of SF6.

– Figure 3d, sweeping overstroke: The moving parts finish their course whereas the cold gas injection continues until the complete opening of the contacts has occurred.

The fractured pole-unit was taken apart and the top and bottom part are shown in Figure 4. The brittle fracture did run its course through most of

Figure 4. Two halves of the fractured pole-unit after it was taken apart.

the circumference of the enclosure, except for about 50 mm which was cut through.

Some of the material was cut from the enclosure for test coupons. It was found that the casting resin was very notch sensitive i.e. a barely visible 1mm deep scratch at the surface reduced the strength by 40%.

Further investigation revealed that the fractured pole-unit had recently been transported to a another location. It is therefore suspected that the enclosure got damaged as a result of transport and handling and that the operators had not noticed this, since the damage was barely visible.

Impact damage which remains unnoticed or is barely visible, can be extremely dangerous in carbon/thermoset composites because a large chunk of the composite's residual compressive strength is being lost.

4.2 *Hardware failure in tilt-rotor aircraft*

The XV-15 tiltrotor aircraft is a promising candidate for future civil and military applications which combine the vertical lift, hover and maneuverability advantages of the helicopter with the greater forward speed of the fixed-wing airplane. One experimental airplane operated in a NASA/Army test program has been utilized to prove the concept and provide design in-

Figure 5. XV-15 in vertical takeoff and forward flight respectively.

formation for later operational aircraft of this type. The XV-15 has helicopter-like rotors for vertical takeoff and landing shown in Figure 5a. Once airborne, the rotors tilt forward to become propellers for cruise flight as shown in Figure 5b. The XV-15 can fly roughly twice as fast and twice as far as a helicopter on an equal amount of fuel; it has achieved top speed of 364 miles per hour and reached 115 miles per hour in the helicopter mode.

The mishap occurred when the aircraft was completing a series of low level passes, under different flight conditions, over an acoustic array. After making six passes the aircraft turned and climbed with nacelles set at 70 deg, RPM at 92% and gear down. At an altitude of 253 m above ground level, there was a loud report heard in the cockpit, followed by greatly increased vibration. At the same time high loads were seen on the ground telemetry. These confirmed that there was a high vibratory condition on the left nacelle. The test team correctly concluded that there was an unbalance on the left rotor.

The pilot immediately turned for the runway at the onset of vibration and the aircraft touched down 82 seconds after the incident.

An immediate inspection of the left rotor revealed that the composite blade cuff had moved outboard approximately 200 mm and, as can be seen in Figure 6, had cracked on both its leading and trailing edges.

Investigation of the evidence revealed that the fiberglass flanges had delaminated, as shown in Figure 7, and were therefore no longer reacting the centrifugal loads. As a result the cuff had started to travel over the blade in the direction of the centrifugal forces. This same travel also caused the leading and trailing edges to split.

Serious errors have been committed repeatedly by analyzing the strong

Figure 6. Cracked blade cuff.

Figure 7. Delaminated fiberglass flanges.

direction of parts and tacitly assuming that no stress will occur in the weak direction.

The loss of structural integrity was due to secondary (interlaminar) tensile stresses, which are not normally analyzed in metal construction. It was found, that even in the most favorable case where the cuff was assembled just right, the ILTS was still 62% of ultimate, a situation which is untenable from a fatigue point of view. Analysis [24] [25] also revealed that the stresses turned out to be rather sensitive for assembly tolerances.

4.3 *X-wing failure*

A NASA Rotor System Research Aircraft (RSRA) has been modified to incorporate an extremely stiff, four-bladed rotor designed to be stopped in flight. The aircraft configuration is shown in Figure 8. The design called for takeoff and low-speed flight in the conventional rotary wing mode. At

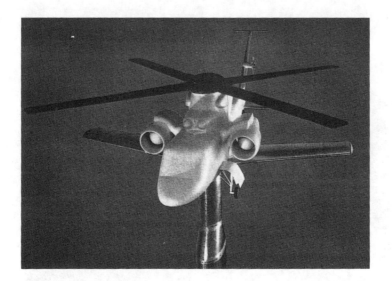

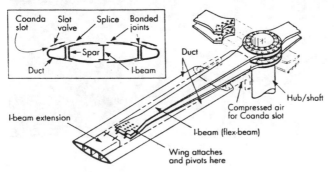

Figure 8. NASA Rotor System Aircraft with 'X' wing.

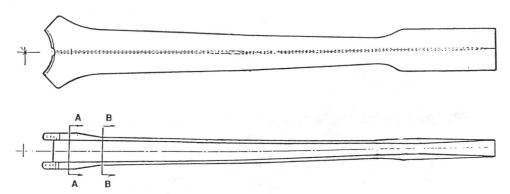

Figure 9. Flexbeam for the wing.

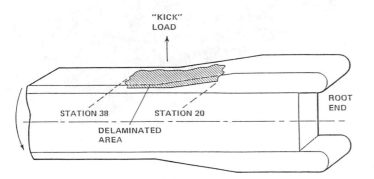

Figure 10. Flange delamination caused by kick loads.

speeds of about 200 mph, the rotor was designed to be stopped and locked in place, converting the RSRA to a fixed wing aircraft with two forward swept and two aft swept wings in a 'X' configuration.

The flexbeam for the wing shown in Figure 9 has thick flanges at the hub in order to provide sufficient material to diffuse the structural loads into the attachment configuration. Plies were internally dropped along the length of the flange to produce a taper. The introduction of these structural disconti-nuities within the laminate produced kick-loads and the associated out-of-plane stresses, which did cause delamination of the composite structure, as shown in Figure 10.

Kim & Dharan [26] developed fracture control plans for composite struc-tures in which delamination fracture is emphasized. It involves materials and process control, proof testing, design aspects, fracture criteria etc. They have given two specific examples; one which applies to the delamination in the flanges of a rib-stiffener and a second for the design of adhesively bonded joints in composite structures.

PROGRAM OBJECTIVE
Develop a Damage Tolerant Composite Wind Tunnel Blade

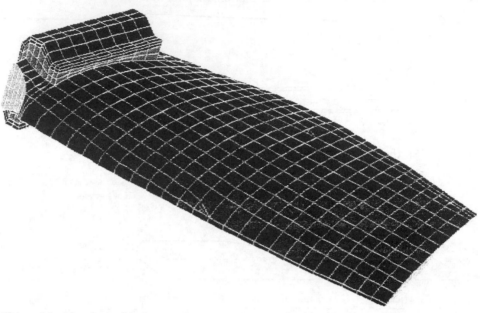

Figure 11. Aluminum blade.

4.4 *Importance of design aspects for safe structural integrity of a wind-tunnel compressor blade*

The structural integrity problems which were mentioned in the two previous examples were due to load introduction problems, in other words they were problems which were directly and indirectly associated with joints.

This last example has been offered in order to illustrate that design for safe structural integrity can be done right.

A good design was needed to demonstrate the feasibility of replacing a set of existing aluminum blades, in a large pressurized high speed wind tunnel, with composites. Figure 11 shows the aluminum blade which is basically machined out of a billet of aluminum. In Figure 12 is the composite blade of which the most critical and important part is the design of the retention.

The long-twisted airfoil shape has been designed as a damage tolerant sandwich panel, the specifics of which will be explained in the lecture. It can be seen that the root was designed to flare-out like a trumpet near the end. This was done in order to gain a mechanical as well as an adhesive retention in the root block. The combination of these two joining methods provided a highly reliable blade-to-root-block connection which upon pro-

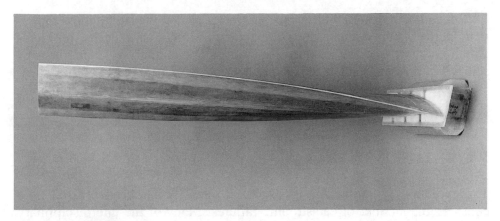

Figure 12. Composite blade.

totype testing demonstrated a safety factor of seven, which was well above the minimum factor of five required to run hardware in the wind tunnel without the benefit of continuous monitoring by strain gages.

Another essential part of the design was driven by damage tolerance requirements. It was found that much of the technology developed in the aerospace needed to be modified in order to be valid for a land-based application [28-33].

Other projects which make use of these joints have been extremely successful and reliable. An advanced version of these joints [34] has been proposed for the 15 story building which was mentioned in Section 3.

One final philosophical comment in closing; when one thinks of American Indians or Polynesians, one thinks of wood, rope, things lashed together or maybe glued. Looking at rafts, Polynesian houses and huts one wonders what would happen if we had to get along without nails, screws, pins, nuts and bolts. Lashing distributes stress over a wide area, whereas nails and other pin fasteners direct the connecting forces to points causing high stress concentration. Stress concentrations threatens the downfall of every piece of dynamically-loaded machinery. In composites we are returning to the old world technology, and to nature (composite is synthetic wood). Fibers are used at concentrated load points by wrapping them around the mating parts like the lashing of poles together. Attaching hardware is eliminated. Just as the Polynesian canoe is carved out of one tree, with no joints or fasteners, so is the optimum rotor blade co-cured in one big lump with no joints whatever and with the fibers trapped while running in the optimum direction to resist the strains (or provide the needed flexibility or stiffness) at each location. Trees, bones and other natural objects operate in this manner and we are trying to re-invent nature in the shape of a wind-tunnel compressor blade, a tail-rotor drive-shaft , a fuselage, etc.

'Education in design with composite materials', is the single most important factor which could reduce the type of structural integrity problems which we discussed in the four case studies. While analysis in engineering science is an important facet of engineering it is clear that synthesis oriented skills such as design have been neglected. Today the top engineering students are the top analysts, not the top designers [35, 36].

5 CONCLUSIONS

The evolution of knowledge in structural integrity has been spurred by a considerable number of catastrophic failures. Many of the interdisciplinary technologies required for the formulation of safe structural integrity programs were developed as a result of these failures. To date the problem of life prediction for metals has not been solved, although one has attacked it for the last 150 years. Some of the current problems were dramatically demonstrated at the Army's Helicopter Fatigue Methodology Specialist Meeting in March 1980, in which various manufacturers, from around the world, presented vastly different conclusions to a simple type of 'canned', text-book-style sample problem.

The cost of fracture in the US alone is $120 billion/year of which $40 billion could be avoided by using current practices.

If there is one thing to be learned from the staggering cost of failure in metallic structures it should be this: 'current best practices are not being used as much as they should because the Structural Integrity Procedures and Methodologies are considered the domain of the 'expert.' To make progress they should be placed in the hands of the people who design and build structures and components.'

The pace of R&D in composite materials and structures is very vigorous. This sets it apart from many older engineering disciplines where, as the research continues, the experts gradually change fields, lose interest, and retire, ...new scientists rediscover the old paradoxes and handily manipulate a few somewhat accepted parameters..., and in this way the generations come and go.

Research proponents and contractors have enthusiastically emphasized the most desirable features of composite materials usage, such as reduction in part count, reduced weight and reduced cost. As the case studies revealed obvious problems such as the liability to split like wood in the weak direction, and to shatter under impact have sometimes been de-emphasized. Serious errors have also been committed repeatedly by designers analyzing the strong direction of the part and tacitly assuming no stress will ever occur in the weak direction.

'Education in design with composite materials', is the single most important factor which could reduce the type of structural integrity problems which we discussed in the four case studies. While analysis in engineering science is an important facet of engineering it is clear that synthesis oriented skills such as design have been neglected. Today the top engineering students are the top analysts, not the top designers [35, 36].

One should approach designs and theories with sufficient caution and healthy skepticism, for the history of science and engineering is littered with failed promises...

REFERENCES

[1]   Petroski, Henry, Failed Promises, American Scientist, Volume 82, January-February 1994, pp. 6-9.

[2]   Bannantine, J.A., Comer, J.J. & Handrock, J.L., *Fundamentals of Metal Fatigue Analysis*, Prentice Hall, 1990.

[3]   Serling, Robert J., The Jet Age-the epic of flight-, Time-Life Books, Alexandria, Virginia, 1982.

[4]   Dempster, Derek, D., *The Tale of the Comet*, David McKay, New York, 1958.

[5]   National Materials Advisory Board, *Assuring Structural Integrity in Army Systems*, National Academy Press, 1985.

[6]   Hansen, A.G., *Damage Tolerance, Aerospace Engineering*, June 1989, pp 19-21.

[7]   Tiffany, C. F., *The Air Force's Changing Philosophy on Structural Safety, Durability and Life Management.*'

[8]   Reifsnider, K., Presentations in AIB-Vinçotte Chair, Thursday March 10, 1995.

[9]   Halpin, J.C., Waddoups, M.E. & Johnson, J.A., *Kinetic Fracture Models and Structural Reliability*, Int. J. Fracture Mechanics, Vol 8, 1972, pp 465-468.

[10] Irving, Clive, *Wide-Body-The Triumph of the 747-*,William Morrow and Company, 1993.

[11] Reed, R.P., Smith, J.H. & Christ B.N., The Economic Effects of Fracture in the United States, U.S. Department of Commerce, National Bureau of Standards, Special Publications, 647-1 & 647-2, 1983.

[12] Faria, L., The Economic Effects of Fracture in Europe, Final Report, Study Contract No 320105, between the European Atomic Energy Community & Stichting voor Toepassing van Materialen (Delft), CEC, 1991.

[13] Dowling, N., *Mechanical Behavior of Materials*, Prentice Hall, 1993.

[14] Time Magazine, August 19, 1991.

[15] Aerospace Composites and Materials, Fall 1989, pp.11.

[16] Baker, A.A. & Jones, R., *Bonded Repair of Aircraft Structures*, Martinus Nijhoff, 1988,

[17] Kaempen, C., 21-st SAMPE workshop on *Tapping the Markets for Industrial Application*, Jan 26, 1995. Sunnyvale, CA.

[18] McErlean, D.P., *Benefits and limitations of Composites in Carrier Based Aircraft*, NASA/FAA/DOD meeting, Nov 1991, Lake Tahoe, Nevada.

[19] Yoda, N., *The Polymer Industry Beyond the Year 2000: Dynamism of world*

trade and economy-business alliance in the Asia pacific region, Journal of Polymer Science, POlymer Symposium 75, 1993, pp 125-136.

[20]   Hiel, C., Personnal communication NASA Code JEF, Jan 1994.

[21]   Fuller, B., *Ideas and Integrities*, Collier Books, 1963.

[22]   Patfoort, G., *Architecturaal Gebruik van Komposieten*, KU Leuven, Studiedag lichtgewicht Komposieten, Mei 1982, pp.64.-73.

[23]   Hiel, C., Personnal communication with Brandt W. Goldsworthy, Feb 1995.

[24]   Hiel, C., Sumich, M. & Chappell, D., *A Curved Beam Test Specimen for Determining the Interlaminar Tensile Strength of a Laminated Composite*, Journal of Composite Materials, vol. 25-July 1991, pp 854-868.

[25]   Hiel, C., *Report to chairman of XV-15/ATB cuff mishap investigation board*, Nov 14, 1991.

[26]   Kim, W.C. & Dharan, C.K.H., *A Fracture Control Plan for Composites*, Engineering fracture Mechanics, 1989.

[27]   Hiel, C., Dittman, D. & Ishai, O., *Composite Sandwich Construction with Syntactic Foam Core-a practical assessment of post-impact damage and residual strength,* Composites, Vol 24. No5, 1993, pp 447-450.

[28]   Hiel, C. & Ishai, O., *Design of highly damage-tolerant sandwich panels*, 37th SAMPE symposium, March 1992, pp. 1228-1242.

[29]   Ishai, O. & Hiel, C., *Damage Tolerance of a Composite Sandwich with Interleaved Foam Core*, J. Composite Technology Res 14, No3, Fall 1992, pp 155-168.

[30]   Hiel, C., & Ishai, O., *Effect of Impact Damage and Open Hole on Compressive Strength of Hybrid Compressive Strength of Composite Skin Laminates*, ASTM Symp on Compression Response of Composite Structures, 16-16 Nov, 1992.

[31]   Ishai, O., Hiel, C. & Luft, M., *Long-Term Hygrothermal Effects on Damage Tolerance of Hybrid Composite Sandwich Panels*, Composites, Vol 26, No 1, 1995, pp 47-55.

[32]   Hiel, C., *Research In Design for Assembly-extending the use of 'snap fit' assembly methods to composite materials*, NASA DDF Paper 7/20/1994

[33]   Kerr, A.D. & Pipes, R.B., *Why we need hands-on Engineering Education*, Technology Review, October 1987, pp 37-42.

[34]   Biot, M.A., *Are we Drowning in Complexity?*, Mechanical Engineering, February 1963, pp.26-27.

[35]   How Things Work-structures, Time-Life Books, 1991.

[36]   Calder, R., After the Seventh Day-the World Man Created-, Simon and Schuster, 1961

[37]   Proceedings of the *Helicopter Fatigue Methodology Specialists' meeting*, March 25-27, 1980 St. Louis, Mo.

APPENDIX 1: ESTIMATED REMAINING SERVICE LIFE USING A FRACTURE ANALYSIS

As an example of the general application of the principles and ideas briefly discussed in the introduction, we will consider an hypothetical pressure vessel and through a rough application of a fracture analysis attempt to estimate the remaining service life of the vessel. This is meant in no-way to be a rigorous approach. Numerous examples of rigorous analyses can be found in the literature particularly as related the petroleum industries [1], rubber articles, including tyres [2], the gas industry [3], in the design of wood structures [4] and in the design and operation of nuclear plants [5].

As Broek states in [6]: 'The time for application of fracture mechanics is NOW! If a fracture occurs, lawyers, in todays litigious society, will be quick to point out that the best available techniques were not used.'

Given

Let us assume the hypothetical pressure vessel is a welded, cylindrical vessel and was constructed in the late 1960's of material meeting the ASTM Standard A212-B. This is an iron base alloy containing 0.35% C and 0.90% Mn. It has a yield strength of 262 MPa , an ultimate strength of 482 MPa, and an elongation of 21%. The vessel is 15.2 m long, and has an interior diameter of 0.91 m and a 100 mm thick wall. It has forged hemispherical end caps which are welded to the body of the vessel. The vessel was designed to contain commercially pure nitrogen at 35 MPa and to be cycled from 35 MPa to 0 psi at an average frequency of 2.5 times daily. The vessel was sucessfully proof tested at 1.25 times its operating pressure in June 1974 and put into service in January 1975. It has operated continuously to date. The maximum applied stress on the vessel is the circumferential stress in the cylinder wall and is equal to 155 MPa at max operating pressure and was 194 MPa during the proof test. To date the vessel has been exposed to approximately 17,562 full pressure cycles.

Note: The formulae which are being used can be looked up in any standard textbook that includes a chapter on fracture mechanics or in [6] where a wealth of practical design information is provided.

Assumptions

The literature was surveyed and no information on the fracture toughness of ASTM A212-B was found. It appears that ASTM A212-B is essentially identical to ASTM A516-Grade 70. Literature indicates that these types of material exhibit a minimum plane strain fracture toughness (K_{IC}) of about 55 MPa $\sqrt{m}$. This agrees with other reported values for similar mild (medium carbon) steels.

Literature was surveyed for subcritical crack growth data. Again, no reliable fatigue crack growth information could be found. However, data on four steels similar to ASTM 212-B were found and are shown in Figure A1. All data appear to follow similar fatigue behaviour with a narrow range of scatter. Therefore, the upper limit of this data was assumed to represent the fatigue crack growth of A212-B and is given by the relation:

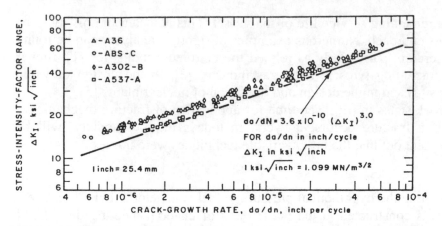

Figure A1. Summary of fatigue-crack-growth data for ferrite-pearlite steels.

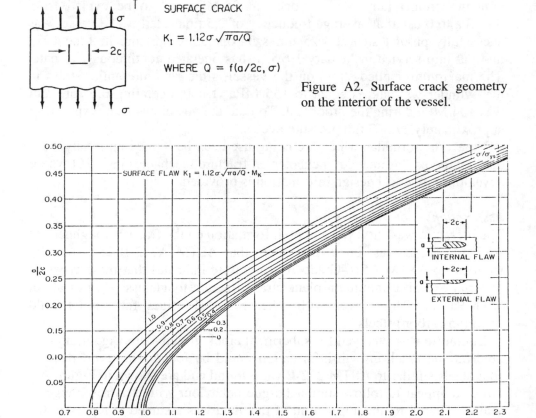

SURFACE CRACK

$$K_I = 1.12\sigma\sqrt{\pi a/Q}$$

WHERE $Q = f(a/2c, \sigma)$

Figure A2. Surface crack geometry on the interior of the vessel.

SURFACE FLAW $K_I = 1.12\sigma\sqrt{\pi a/Q} \cdot M_K$

Figure A3. Crack-shape parameter for a surface flaw.

$$\frac{\mathrm{d}\,a}{\mathrm{d}\,N} = 6.88 \; 10^{-12} (\Delta K_I)^{3.0}$$

where: $\mathrm{d}a/\mathrm{d}N$ is the fatigue crack growth rate in m/cycle and ΔK_I is the alternating stress intensity.

The most probable crack location to be considered was assumed to be a surface crack on the interior of the vessel as schematically shown in Figure A2. Additionally, the crack was assumed to have an aspect ratio (a/2c) of 0.20. This would result in a value of approximately 1.25 for the crack shape parameter (Fig. A3) Therefore the functional relationship between stress intensity, applied stress, and crack length can be expressed as:

$$K_I = 1.1 \, M_K \sigma \sqrt{\frac{\pi a}{1.25}}$$

$$K_I = 1.75 \, M_K \sigma \sqrt{a}$$

where: M_K is the magnification factor.

Question
If failure of the vessel was to occur, would it be by a leak through the vessel wall or by bursting of the vessel?

Answer
From the equation, assuming $M_K =$

$$a_c^o = \left(\frac{K_{IC}}{1.75\sigma_p}\right)^2 = \left(\frac{55}{1.75\;155}\right)^2$$

$$a_c^o = 4.1\,10^{-2} \text{ m}$$

$$a_c^o = 41 \text{ mm}$$

At the maximum operating stress the critical crack length for rapid unstable fracture is much less than the wall thickness of the vessel (100 mm). Therefore, the vessel will fail by a burst rather than by a leak.

Question
What is the maximum existing crack size that was established by the proof test?

Answer

$$a_c^p = \left(\frac{K_{IC}}{1.75\sigma_p}\right)^2 = \left(\frac{55}{1.75\;193}\right)^2$$

$$a_c^P = 2.6 \ 10^{-2} \ \text{m}$$

$$a_c^P = 26 \ \text{mm}$$

Question
How much would the maximum size existing crack have grown during operation of the vessel to date?

Answer
The ΔK_I at a crack lenght of 26 mm for a full pressure cycle (34 to 0 MPa) would be

$$\Delta K_I = 1.75 \ \Delta\sigma\sqrt{a} = 1.75 \times 155 \ \sqrt{26 \ 10^{-3}}$$

$$\Delta K_I = 44 \ \text{Mpa} \sqrt{m}$$

Therefore, the fatigue crack growth rate at this value of ΔK_I would be

$$\frac{da}{dN} = 6.88 \ 10^{-12} \ (\Delta_{KI})^3 = 6.88 \ 10^{-12} \ (45)^3$$

$$\frac{da}{dN} = 627 \ 10^{-9} \left(\frac{m}{\text{cycle}} \right)$$

$$\frac{da}{dN} = 627 \ 10^{-3} \left(\frac{mm}{\text{cycle}} \right)$$

In order to estimate the total crack growth in 17,562 cycles, growth must be broken into increments. We have chosen increments a of 1.5 mm. Therefore at a d*a*/d*N* equal to mm/cycle, it would take 2,392 full pressure cycles for the crack to grow 1.5 mm. The new value of a would be 27.5 mm, ΔK_I would then equal 45 Mpa... The above process is repeated until the total number of cycles has been accounted for. Using this approximate method, the crack is estimated to have grown from an initial size of 26 to 38.6 mm in 17.562 full pressure cycles.

Question
Is the vessel safe for continued operation?

Answer
During the operating life of the vessel, the crack is estimated to have grown 85% of the distance between its maximum initial size of 26 mm established by the proof stress and the critical crack size of 41 mm established by the operating stress. The estimated present crack length is close to the point where the vessel will fail and the vessel should thus be removed from service.

Question
Can this vessel be recertified for service?

Answer
Obviously, if in fact a crack does exist in the vessel and is approaching the critical crack length, continued service at the designated operating stress would result in failure. However, if the vessel were to be derated to say a maximum operating pressure of 27.5 MPa from the original design pressure of 35 MPa, the stress in the vessel would be reduced from 155 to 124 MPa and the critical crack length would be increased from 40 to 63 mm the vessel then could operate safely for a number of additional years. The estimated remaining life could be calculated as above based upon a maximum existing crack size based upon the previous operating stress.

Alternatively, the vessel could be operated at the design pressure only if the vessel were successfully proof tested. Again, the proof test would establish a maximum existing crack size sufficiently less than the critical crack size to allow the vessel to continue its safe operation.

REFERENCES

[1] Cotton, H.C., *Experience in the Petroleum Industries*, pp 179 in Fracture Mechanics in Desogn and Service. The Royal Society of London, 1981.

[2] Breidenbach, R.F. & Lake, G.J., *Application of Fracture Mechanics to Rubber Articles, Including Tyres*. pp 189 in Fracture Mechanics in Design and Service. The Royal Society of London, 1981.

[3] Fearnehough, G.D. & Greig, J.M., *Experience in the Gas Industry*, pp 203 in Fracture Mechanics in Design and Service. The royal Society of London, 1981.

[4] Barrett, J.D., *Fracture Mechanics and the Design of Wood Structures*, pp 217 in Fracture Mechanics in Design and Service. The Royal Society of London, 1981.

[5] Nichols, R.W. & Cowan, A., *Fracture Mechanics as an Aid to Design and Operation of Nuclear Plant*, pp 227 in Fracture Mechanics in design and Service. The Royal Society of London, 1981.

[6] Broek, D., *The practical use of fracture mechanics*, Kluwer Academic Publishers, 1989.

Design aspects related to the durability of composite structural components

G. Jeronimidis
Department of Engineering, Reading University, Reading, UK

ABSTRACT: In the past few years composite materials technology has moved steadily from the aerospace sector, in which they have evolved, into general engineering applications. There has been progress but not as rapid as one might have expected. Problems related to durability and long-term reliability of composite structural components will play a major role in their acceptance and success. Economics are also of main concern in volume applications of composites. A great deal of creative thinking is needed at the conceptual design stages to ensure that cost-effective performance levels are achieved over acceptable time-scales. In many industries benefits from composites are more likely to come from balance of properties and tailored solutions than from ultimate levels of performance.

1 COMPOSITES IN NON-AEROSPACE APPLICATIONS

Polymer-based fibre-reinforced materials are the most versatile and useful group of composites for general engineering applications. In some instances (glass fibres, polyester and epoxy resins, for example) the raw materials costs are not significantly higher than those of more traditional engineering materials. Even when the price differential is penalising superficially composite materials (aramid and carbon fibres), current processing and fabrication technologies are well developed to offset higher materials costs by efficient components manufacture. Pultrusion, resin transfer moulding as well as highly-developed filament winding, are particularly suited to provide design flexibility and reproducible high quality (Gay 1989, Gibson 1994).

Several applications sectors have used composites for many years, others are just beginning. A list of the major non-aerospace opportunities for durable composite structures is given in Table 1. With few exceptions, most of the applications listed in the table need appropriate levels of performance, cost and durability to compete effectively with established designs.

173

Table 1. Non-aerospace applications sectors for structural composites.

Road and rail transport	Shell structures, springs, suspensions, drive shafts
Energy	Wind turbines, flywheels
Marine	Leisure boats, rescue boats, commercial transport
Oil	Extraction platforms, tethers
Medical	Internal & external prostheses, hospital equipment
Sport	Rackets, racing cars, competition boats, skis, ...
Civil engineering	Water purification tanks, cladding, architectural elements, reinforced concrete
Security	Blast, fragment and bullet protection

2 KEY ELEMENTS IN COMPOSITE DESIGN

Continuing advances in the development of fibres and matrices are not reflected in the rate of application of composites, especially in non-aerospace industries. The full potential of composite systems is yet to be realised but far too much effort is perhaps devoted to the development of new, increased-performance fibres and resins rather than to the extraction of maximum benefits from existing materials. All too often there is a tendency to associate high performance with *composite properties* rather than *good design*.

This is due to a large extent to design approaches which impose severe limitations on creativity and full exploitation of composite concepts. The early 'substitution' strategy of replacing existing metal components with composite ones may at best be defined as rudimentary. It is difficult to see how a design based on isotropic, homogeneous materials can really benefit from switching to anisotropic and heterogeneous ones. Similarly, 'cut and try' strategies are unlikely to stimulate the rational conceptualisation and integration which are needed in composite design.

There are three major factors which have limited, and still limit, the use of structural composites:
 – Lack of confidence in the design process;
 – Misconceptions about true *total cost* of composite solutions;
 – Concerns about long term durability.

When dealing with structural composite applications, comparing only materials costs rather than total solution costs will often result in materials selection processes which do not take into account the additional added value that composites can provide. Identifying where this value lies is crucial.

In aerospace, structural weight saving is a design driver which helps to give emphasise the importance of specific stiffness and specific strength which fibre-reinforced materials can offer. The weight saved in using composites in aircraft components is easily translatable into increased payloads, fuel economy, etc., but for most of the applications listed in Table 1 there is not a clear-cut relationship between weight saved and performance.

In fact, in the majority of general engineering applications of composites, weight saving is of minor importance, and often not relevant at all.

This needs to be appreciated in order to redirect the design strategy away from minimum weight solutions, towards concepts where other design-properties combinations, such as durability for example, represent the real advantages in the use of these materials.

Durability aspects are particularly relevant because the polymer matrices used are generally considered the weak link of the composite structures. Their mechanical properties are much lower than those of the fibres and they are also the component which are exposed first to water, chemicals, etc. Understanding how potentially damaging combinations of loads (static, dynamic, fatigue) and environments (water, organic solvents, temperature) affect the service life of composite components is clearly of major importance in increasing the level of confidence of composite systems (Cardon & Verchery 1991).

Equally important is to develop effective methodologies (testing, analysis, simulation) capable of taking into account these limitations at the early stages of the design process. Compromises and optimisation for multiple requirements should be assessed and implemented at the earliest possible opportunity in order to integrate as fully as possible materials, shapes and fabrication. Predictive models of damage initiation, accumulation and propagation can help considerably in selecting global and local fibre orientations, load paths, smooth transitions, etc. so as to minimise the matrix degradation processes.

Composite failures where the fibrous phase is still virtually intact when failure takes place cannot be considered 'robust' solutions. Delamination failures in composite laminates are a typical example of matrix-dominated fracture mechanism which lead to severe degradation of performance (bending stiffness, buckling resistance, compressive strength) with the fibres still perfectly functional. Significant improvements in minimising or eliminating this kind of problem are more likely to come from design solutions than from marginal, and often costly improvements in the materials themselves.

Some of the key elements which need to be considered for maximum-benefit, cost-effective composite systems are outlined below and will be discussed in more detail in the examples presented later on:

Conceptual designs – creativity – total solutions:
   – Identify composite advantages in relation to application: density, anisotropy, specific properties, fatigue life, corrosion resistance ....;
   – Identify uniqueness of composite solution in relation to specification;

- Identify constraints inherited from established non-composite designs, regulations, etc.; assess limitations imposed by constraints;
- Tailored solutions, integration of functions; integration of components;
- Materials selection (fibres, matrices, hybrids, selective reinforcement);
- Assessment of failure modes and damage mechanisms (fibre-dominated, matrix-dominated);

Integration of conceptual designs with fabrication routes.

Development of test programmes to validate choices:
- Materials testing; component testing;
- Assessment of failure modes and damage mechanisms (fibre-dominated matrix-dominated);
- Multiaxial loading;
- Durability testing; interactions loads-environment;
- Design allowables.

Detailed design – design tools:
- Fibre architectures; stress analysis;
- Load-transfer aspects (joints, etc.);
- Predictions of performance levels (analysis, modelling);
- Damage and durability models;
- Failure modes: structural integrity, safety aspects.

Quality control strategies and techniques.

Inspection, NDT, repair issues during life cycle.

3 DESIGN STRATEGIES FOR LONG-TERM DURABILITY AND RELIABILITY

Durability and reliability of performance in composite structures are critically dependent on the detailed design process which finalises the component. It is at this stage that the initial concepts need to be refined and optimised in the light of the available information (testing, analysis, etc.). Weaknesses and potential trouble areas should also be anticipated.

The application of some general design principles, particularly important with composites, can provide a check on the overall 'balance' of the adopted solution. For example:

Shape and geometry
Are the shape and geometry simple and compatible with the chosen fibre directions for primary loading(s)?

Are changes of radii, thickness, etc., sufficiently smooth to avoid local discontinuities and associated stress concentrations and/or parasite loads?

Are changes of thickness *really* needed? The additional material cost and weight penalty of having constant thickness may be trivial in comparison to the design complications involved in dropping plies, for example, and ensuring an effective and reliable load transfer in the ply-drop region.

If thickness changes are unavoidable, can they be obtained maintaining the integrity and continuity of the fibre system? Use of local core elements, for example.

Secondary stresses

Have *all possible* sources of secondary loading been identified in relation to real component function?

Are these loads carried primarily by the matrix? This is often the case and it must be remembered that five out of six basic composite failure modes are matrix-dominated and do not make any use of the fibre strength available: axial compression, transverse tension and compression, in-plane shear, delamination.

Do the composite components *conform* to the type of assumptions made? Is the composite structure a true 2-dimensional object, which can be analysed with simple laminate theory (Jones 1975, Powell 1994)?

Are through-the-thickness shear and normal stresses significant (thick sections, changes of thickness or curvatures) especially in components loaded in bending? Are these stresses carried by the matrix? Transverse shear stresses resulting from the shear force introduced by transverse loads can easily become the dominant factor in both stiffness and strength controlled applications.

If significant interlaminar shear and transverse tensile normal stresses are present, can they be reduced by re-design within the primary functional requirements? If not, can the relevant local strength where these stresses are dominant be increased over and above what the matrix alone can provide? This may be possible by using different intermediate fibre structures, weaves, for example, where the fibres contribute somewhat to carry these stresses because of their local orientation, or by stitching.

Have thermal stresses be taken into account? The problem of thermal stresses is particularly relevant to polymer resin systems which are processed at high temperatures, both thermoplastic and thermosetting. Residual stresses in conventional epoxy-based laminates, especially those based on carbon or aramid fibres, can easily reach 30% of the transverse tensile strength of the constituent plies and lead to premature transverse cracking. This, in turn, can trigger delamination failures or provide pathways for the ingress of water and chemicals with detrimental results for durability and long-term reliability. Both mechanical and thermal cycling can lead to rapid deterioration in the presence of residual thermal stresses (Jeronimidis & Parkyn 1988).

Dimensional stability

Changes of shape due to build-up of residual stresses during cooling from the processing temperature can be very significant, especially in components of large dimensions. This affects in particular unsymmetrical laminate constructions and although the development of residual curvatures alleviates to some extent the magnitude of the thermal stresses, composite substructures which have to be joined to other rigid elements may need to be deformed, and hence pre-loaded, before assembly (Hyer 1983).

Holes and fasteners

In composites design one should try to eliminate the need for post-machined holes in structural components. These systems are not particularly suited to carry bearing stresses which act generally in one of the weak directions of the materials. In addition, the free edges which are created by the provision of holes for fasteners give rise to localised three-dimensional stress-states which are often responsible for premature delamination (Pagano 1989).

All to often the 'substitution' approach mentioned earlier leads to conflicts of this kind; the design was not conceived from the start with composite materials in mind. Holes and mechanical fasteners, which work well with metals, are very inefficient and potential sources of long-term degradation processes. One must not forget that the fibres are the major load-bearing in the system and disrupting fibre continuity, and hence stress continuity, by cutting fibres locally is asking for trouble.

Differences between the coefficients of thermal expansion of the composite and the fastener/sub-structure can lead to rapid deterioration of the joint, especially in fatigue conditions and in aggressive environments.

The need for mechanical joints should be assessed very carefully in order to eliminate them as much as possible. This means that the composite solution to a component design problem ought to be developed as an integral part of the overall structural design approach and not in isolation. If composites are wanted, they have to be there from the start.

Bonded joints

Bonded joints are generally a better alternative for composites than mechanical joints because polymer-based adhesives have mechanical and thermal properties which are closer to those of the matrix than metals. In spite of this, joints are critical areas where load transfer takes place and it is important to think about the possibility of maintaining the integrity of the fibre system by overwrapping, looping, etc. When possible to implement, this approach has the advantage of utilising in full the tensile strength of the fibres and relying directly on tensile or bending stresses to transmit loads rather than shear stresses at the joint interfaces.

4 EXAMPLES OF INTEGRATED APPROACHES TO COMPOSITES DESIGN

In this section three examples of structural design with composites will be illustrated. Detailed analysis and calculations have been omitted for clarity in order to emphasise the thinking processes which have led to the composite solutions. Additional information can be found in the references.

4.1 *Composite drive shaft*

The free, unsupported length between bearings of metal drive or prop shafts for torque transmission in automotive applications and power plants is often limited by the critical whirling speed of the rotating component. This is a dynamic instability which depends on materials properties, Young's modulus and density, as well as on geometrical variables such as unsupported length and shaft diameter.

Figure 1 shows the exaggerated shape taken by of a tubular shaft during whirling. It can be shown that, for a given unsupported length, the critical rotational speed associated with the instability is proportional only to the $(E/\rho)^{1/2}$, where E is the Young's modulus of the material and ρ its density.

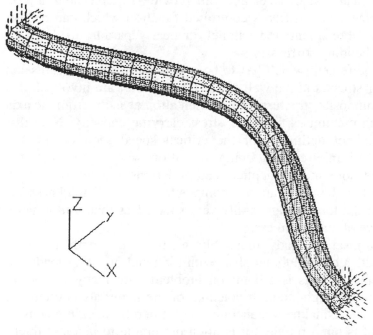

Figure 1. Deflected shape of a tubular drive shaft corresponding to whirling instability.

No modification of size of the circular cross-section or of the wall thickness of the tube has any effect. The shear strength of the steels used in this kind of application is generally more than sufficient for the transmission of the power. At any rate, since the critical speed is limiting and since it does not depend on thickness, this dimension can be adjusted freely to keep the shear stresses within specified limits.

The speed limitation imposed by whirling can be offset only by shortening the unsupported length of the shaft and hence by the addition of intermediate bearings. The cost associated with the bearings is significant in terms of additional components, fabrication costs and reliability. The bearings are also a source of additional noise which may require noise-suppression measures, with associated costs.

It is important to realise that in order to increase critical speeds for a given unsupported length, or increase unsupported length for a given speed, the *only* viable alternatives are materials with a higher specific modulus than steel or aluminium. It is not a question of absolute mass reduction; a successful composite solution may well give a reduction in mass as well but it is unlikely to be significant or important. The real advantage of a possible composite solution is in the increase of unsupported length with elimination of the bearings and noise sources (Pasquier 1987).

Both carbon and aramid fibre reinforced composites, at 50-60% volume fraction of fibres, have specific moduli in the fibre direction which are significantly higher than that of steel and can provide suitable alternatives. This first requirement is a stiffness controlled function which can be met with a composite fibre orientation parallel, or nearly parallel to the shaft axis to maximise bending stiffness.

The second requirement which has to be met is the torque transmission itself. Since shear stresses in the wall of the tubular shaft are involved, this imposes on the composite structure a fibre orientation at ± 45° from the axis of the shaft which maximises the shear stress carrying capacity. Note that this orientation is not optimum for the critical speed requirement; the Young's modulus of unidirectional composites drops very rapidly with the angle between the fibre and the applied load directions. Owing to the high tensile strength of carbon and aramid composites, the wall thickness required to transmit the torque is generally very small, less than a millimetre for the power drive of a medium car.

Thickness optimisation leads inevitably to problems with torsional structural instabilities of the tubular shaft with torsional buckling loads being the limiting factor. This is a common problem with many composite applications where the high strength potential of the materials is often not fully available because thickness reductions are limited by buckling considerations. If weight is not a problem it is often preferable to increase thick-

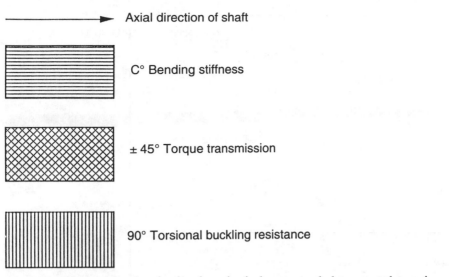

→ Axial direction of shaft

C° Bending stiffness

± 45° Torque transmission

90° Torsional buckling resistance

Figure 2. Fibre orientation in the three basic layers needed to meet the various stiffness, strength and stability requirements in a composite drive shaft.

ness rather than complicate the design and fabrication of relatively thin elements with stiffeners, stringers, etc.

In the case of the drive shaft, torsional buckling resistance can be increased by stiffening uniformly the tube in the circumferential direction, i.e. by adding layers with a fibre orientation of 90° with respect to the shaft axis.

These basic considerations suggest that the composite drive shaft needs three basic layers at different orientations to meet the three basic specifications. These are shown in Figure 2.

A solution combining the three basic layers is possible and the thickness of the various layers can be optimised starting from a simple design where each layer performs only one specific function. Contributions of each layer to the functions of the other two can be calculated to arrive at a construction where each layer performs its primary function but contributes also to those of the others. Further refinements can explore the advantages of different sequences of the three basic layers, from outside in.

Of course, it is desirable to look for a 'single angle' solution which would make the fabrication easier (one setting for filament winding instead of three) and eliminate the difficulty of winding at 0°, or close to this angle. It would also allow for easy integration of the metal end-fittings to get the loads in and out which can be integrally bonded during the filament winding process. Other alternatives involve adhesive bonding of the end fittings after fabrication of the composite tube.

Figure 3 shows a finite elements search for a single angle solution which

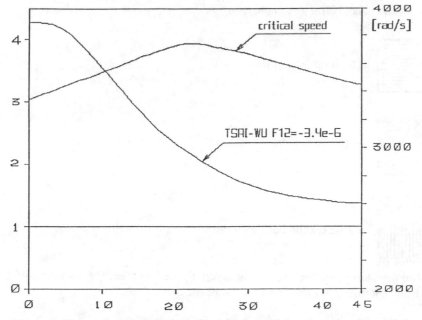

Figure 3. Finite elements analysis for 'single angle' composite drive shaft.

satisfies simultaneously the bending stiffness and shear strength requirements for a drive shaft about 1 m long and with a diameter of 60 mm. The
material used is a high strength-carbon epoxy. The strength assessment has
been carried out using a Tsai-Wu failure criterion.

It is interesting to note that the critical whiling speed for the given length
is maximum for a fibre angle $\theta = \pm 22°$, which is not predictable from first
principles. At that angle the strength criterion is also satisfied (> 1). Further
calculations have shown that the stability requirements are also met.

Note also that the option of aramid fibres-epoxy, which is possible on
specific stiffness considerations will not meet the strength criterion because
the extremely low compressive strength of aramid fibres and aramid-based
composites limits the load-carrying capacity in torsion for one or the other
of the two layers, $+ 45°$ or $- 45°$, depending on the direction of the applied
torque.

4.2 'Smart' composite wind turbine blades

A major problem in wind power generation is the requirement to limit the
overspeed of the rotor at high wind velocity to prevent total structural failure when the induced centrifugal forces on the blades exceed their design
limits. Making the blades stronger is not really feasible in view of the additional weight penalty. Also, extreme gust conditions which may lead to

catastrophic failure are comparatively rare events in the 20 years expected life of the turbine. However, it is a safety critical condition which, even with the small probability associated with it, needs to be taken into account.

Various methods of rotor speed limitation have been developed and tried in the past 20 years. They have been reviewed by Karaolis (1989). Some of them rely on traditional pitch control of the whole blade or part of the blade through gearboxes, shafts and mechanical energy inputs. Rotational speed sensors are also needed to detect overspeed conditions and trigger the change of angle of attack of the blades, towards feather or stall conditions, via suitable actuation. These solutions can be defined as 'active' control. They have the additional advantage of allowing continuous optimisation of blade configuration in relation to wind speed and power generation.

Other solutions are specifically designed to prevent overspeed only. Air brakes, parachutes and similar devices are triggered in response to a rotational speed signal, preventing the destruction of the rotor. These are semi-active devices which need generally to be reset if actuated.

A third group of solutions, passive control, has been used in some design. They are based on aeroelastic concepts where the increase in rotor speed itself forces the blades to rotate and alter their angle of attack.

Composite constructions offer yet another possibility where the passive response of the blade is integrated in the blade structure itself, exploiting coupled deformation phenomena which are typical of certain classes of composite laminates. The advantages of 'smart' blades are:

– Simplicity of the design (no need for sensors, actuators, etc.);

– Reliability and reversibility of the response (which does not depend on devices which may themselves fail to detect and/or actuate);

– Possibility of long-term durability with minimum maintenance through robust designs with comparatively few critical parts such as joints, bearings, etc.;

– Gradual and continuous self-limiting rotational speed at any wind velocity.

The basic design principle consists in coupling the stretching of the blade due centrifugal forces to twisting of the blade section about the blade axis. This is achieved by making the upper and lower surface skins of the blade sandwich structure deform as an antisymmetrical laminate; the fibre orientation on the upper surface is $+\theta$ with respect to the blade axis, that of the lower surface $-\theta$. This fibre arrangement can easily by obtained by filament winding, for example. The mechanism of induced twist under axial tension for antisymmetrical laminates is illustrated in Figure 4a and its translation into a blade section is shown in Figure 4b.

In this design, because the fibres are oriented at an angle θ from the major centrifugal loads, the matrix is deformed in shear; its long term fatigue performance and durability are critical for successful long-term operation.

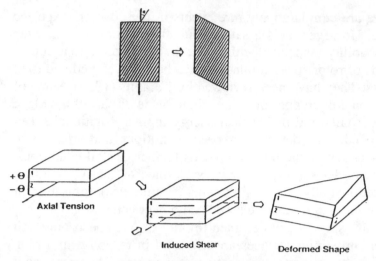

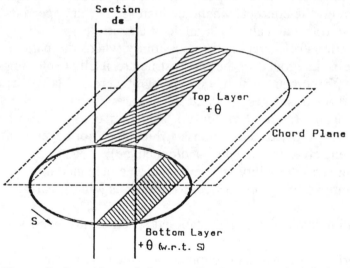

Figure 4a. Basic mechanism of stretching-twisting coupling in composite laminates.

Figure 4b. Stretching-twisting coupling in a composite wind turbine blade section.

Performance assessment needs to be done on representative elements of the composite construction, using the materials and fibre orientations which will be used in practice. Fatigue data obtained from the same materials combinations, but at different fibre orientations than those specified for the stretching-twisting effect, cannot be used reliably since, to all intents and purposes, every fibre orientation represents in practice a different material. This highlights one particular problem of composites design, i.e. the lack of reliable predictive methods to assess fatigue at any fibre orientation from data information obtained from other orientations.

Various materials have been used to quantify the effect for a range of wind turbine aerofoil sections and dimensions. The following general conclusions can be drawn (all the chosen materials have been standardised at 50% volume fraction of fibres):

– The optimum fibre angle needed to maximise the stretching-coupling effect depends on the materials. It has been found to be 18° for glass-epoxy, 10° for aramid-epoxy, and 12° for carbon-epoxy;

– The magnitude of the effect, at the optimum angle, is also strongly dependent on materials properties. Aramid, carbon and glass give, for the same blade geometry, angles of twist in the approximate ratios 3:2:1, respectively;

– Off-axis oriented unidirectional skins give the maximum effect but they are not suitable in practice owing to the very low transverse tensile strength of the plies.

– Unbalanced orthogonal weave fibre structures for the skins have been found satisfactory from both the coupling and strength points of view. The optimum angle position shifts marginally, by a 10° or so, and the magnitude of the coupling decreases because of the counteracting effect of the two principal fibre orientations, but there is still sufficient induced twist for further design and practical implementation.

– The major durability aspect which needs further work is the fatigue performance of the woven laminate structures, oriented at the optimum coupling angles with respect to the centrifugal forces. In extreme conditions, the shear strains can reach values of 0.6-0.8%; the corresponding shear stresses are borne mostly by the matrix which may fatigue in shear.

The design concepts discussed above have been finalised and implemented on a small experimental 3-bladed wind turbine with a blade length of about 1 m. Experimental results have confirmed the applicability and advantages of 'smart' composite designs for this kind of application. Figure 5 shows the relationship between wind speed and rotor speed for the experimental turbine. The straight line corresponds to the response of an uncoupled design, with rotor speed increasing linearly with wind velocity. The other two curves represent the predicted and measured response of the rotor with 'smart' composite blades. The rotors rotational speed increases initially with wind velocity, but levels off at a constant plateau level thereafter. The experimental measurements, obtained in real field conditions, are in good agreement with the design predictions (Karaolis et al. 1988, Jeronimidis et al. 1991).

In many cases, it is not necessary to have stretching-twisting coupling implemented over the whole length of the blade. From an aerodynamic

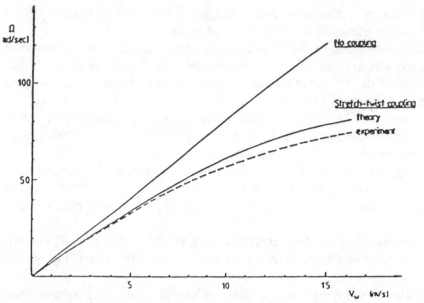

Figure 5. Rotor speed versus wind speed for coupled and uncoupled composite wind turbine blades. Theory and experimental results.

point of view, the first 20% of the blade length from the tip is the more 'active' and where the coupling effect is more beneficial. Recent studies have shown that it is possible to design composite wind turbine blades with 'smart' tips only, leaving the rest of the blade to be optimised for other loads, mainly bending moments associated with flapping, which need axial fibre orientation, and twisting moments which need fibre directions at ± 45° from the blade axis.

Real applications based on these concepts are particularly relevant to medium-large wind turbine rotors, in the range 10-20 m diameter.. In these systems the additional cost of sensing and actuation control devices, which may have to function reliably, but only occasionally, over long periods of time, is a significant proportion of the generator cost. There is also to consider the additional expenses associated with maintenance and testing of the complete active control mechanisms, especially in remote areas where the machines are generally located. All these factors make 'smart' composite designs for blades real alternative solutions, provided that acceptable durability levels can be achieved.

4.3 *Composite springs for road transport applications*

Composite materials are particularly suited for spring applications owing to the relatively high failure strains of the fibres, within linear-elastic range.

Springs are energy-storage devices which need high working strains for efficiency. Glass fibres are better than carbon in this respect because the relatively low modulus of glass allows higher deflections for a given load and hence higher strains.

The use of glass-fibre composite springs in cars, small vans and even large trucks is being increasingly explored because of additional benefits provided by composites, corrosion resistance and, in some cases, weight saving (Lo et al. 1986, Harris 1990). The question of mass reduction needs to be put into context because its importance depends on the specific application.

In cars and small vans, weight saving is not a major issue; additional benefits are wanted and are possible but only at cost-effective levels. In trucks, on the other hand, the total potential weight saving for a multi-axle trailer can be estimated at 350-400 kg. This leads to increased payloads and hence to a clearly defined advantage of composites over metals, even at a higher cost. In recent years, concern about road-friendliness of suspensions systems in relation to road damage has stimulated a renewed interest in composites, mainly glass fibre-based, which offer unique alternatives to metals in this respect (Gyenes 1992).

For light vehicles, the added value of a composite spring system is less clear-cut and often also not easily measurable. Exceptional fatigue life and durability have been demonstrated in some commercial vehicles, with glass-fibres composite springs showing very little signs of damage even after 300,000 miles in service in one year. This is an obvious benefit which translates into reduced maintenance, down-time, etc. There are also safety aspects to consider; when metal springs fail, generally after fatigue crack growth, they do so in a sudden manner owing to the very brittle nature of the hard steels used. Failure will occur without warning whereas composite springs show a more gradual deterioration of properties, often associated with a decrease in stiffness which can act as a warning.

Figure 6 shows the basic geometry and loads for various types of vehicle springs. Depending on the design of the vehicle, not all the loads are reacted by the spring. The major load components which are however common to all springs in all types of vehicles are the vertical load, due to the

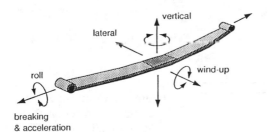

Figure 6. Typical shape and loads for road vehicle springs.

mass of the vehicle plus payload, and the lateral loads due to centrifugal forces in cornering.

These two loads require different response from the spring. The vertical stiffness often needs to be relatively low to provide comfortable ride, whereas the lateral stiffness needs to be high to prevent undesirable sideways motion and roll of the vehicle. Steel springs, owing to the isotropy of the material, cannot satisfy simultaneously both requirements. By the time the lateral stiffness is sorted out, the vertical stiffness is higher than the desirable target value. Shape and thickness profiles can alter this to some extent but they can be costly and, at any rate, the allowable stresses often limit the degree of design flexibility in this respect.

Composite springs offer design solutions where both stiffnesses can be tailored to the specific needs. As well as shape and thickness profiles, fibre volume fractions and fibre orientation are additional design variables which can be used to introduce multifunctionality. A programme of work at Reading University is currently exploring design, testing and fatigue performance of several glass-fibre composite spring types, each tailored to a specific set of functions.

In addition to the advantages outlined above, composite springs can reduce significantly the number of components needed in the suspension system by integrating appropriate elements at the design and fabrication stages. An example of this is shown in Figures 7 and 8 which illustrate, respectively, a basic composite spring design together with the integral eye-end needed to attach the spring to the vehicle. The eye-end must carry horizontal loads associated with acceleration and braking. The elimination of metal brackets has been done by looping the fibre preforms around the bush

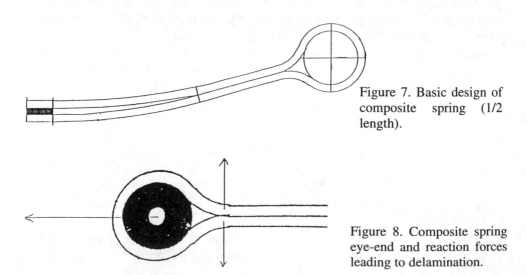

Figure 7. Basic design of composite spring (1/2 length).

Figure 8. Composite spring eye-end and reaction forces leading to delamination.

insert which can be moulded with the whole spring by Reaction Transfer Moulding. The axial load on the spring, reacted by the attachment point, tends to delaminate the spring along its central plane because of the transverse tensile components of the load. This load cannot be carried by the matrix alone and local stitching near the eye end has shown a great deal of promise in this respect.

5 CONCLUSIONS

The potential for polymer-based composite materials in many non-aerospace applications is full of promises which, once demonstrated, ought to place composites among the favourite materials for engineering design. Their versatility and flexibility should be seen as an important asset rather than an additional burden. Their costs, often higher than traditional materials, can be offset by better designs, additional performance, unique capabilities and tailored solutions. Fabrication methods for complex shapes with single or multiple curvatures have become available, Resin Transfer Moulding in particular, and extremely cost-effective. Analytical and modelling tools for predicting performance have also been refined in the past few years. In many ways, most of the necessary elements for the development and implementation of successful designs in composites are in place.

As yet, however, this promising starting point has not been reflected in significant increases in the number or type of applications for which composites are particularly suited. Lack of confidence and lack of proper composite engineering education are perhaps some of the reasons for the slow pace of progress. The two are obviously related since only increased confidence in the designers to tackle engineering problems with these materials can lead to greater use.

Concerns about durability and reliability aspects are also responsible for the perceived lack of confidence. Greater effort is perhaps needed in these areas more than in any other to increase our understanding of the complex interplay between materials, structures, anisotropy, heterogeneity, environment, etc. which are at the root of damage initiation, progression and deterioration of performance. Being able to make good predictions of durability can only help to establish confidence. In parallel there is also a need for design concepts which can exploit as fully as possible the potential of composites systems.

REFERENCES

Cardon, A.H. & G. Verchery (eds) 1991. Durability of polymer based composite systems for structural applications. London: Elsevier Applied Science.

Gay, D. 1989. *Matériaux composites*. Paris: Hermes.

Gibson, R.F. 1994. Principles of composite materials mechanics. New York: McGraw-Hill.

Gyenes, L. 1992. Dynamic pavement loads and tests of road-friendliness for heavy vehicle suspension. *3rd Int. Symp. on Heavy Vehicle Weights and Dimensions, 28th June-2 July 1992*. Cambridge.

Harris, L.R. 1990. *Composite leaf spring design*. London: GKN Technology.

Hyer, M.W. 1983. Non-linear effects of elastic coupling in unsymmetric laminates. In: Z. Hashin & C.T. Herakovitch (eds), *Mechanics of composite materials: recent advances*. Oxford: Pergamon.

Jeronimidis, G. & A.T. Parkyn 1988. Residual stresses in carbon fibre-thermoplastic matrix laminates. *J. Composite Materials* 22:401-415.

Jeronimidis, G., Karaolis, N.M. & P.J. Musgrove 1991. Power control of wind turbine blades through structural design. In: G.K. Haritos (ed.). *Smart materials and structures*. AD-24, AMD-123: 189-202. New York: ASME.

Jones, R.M. 1975. *Mechanics of composite materials*. Tokyo: Mc Graw-Hill Kogakusha.

Karaolis, N.M., Musgrove, P.J. & G. Jeronimidis 1988. Active and passive aerodynamic power control using asymmetric fibre-reinforced laminates for wind turbine blades. D.J. Milborrow (ed.), *Proc. 10th BWEA Conference, 22-24 March 1988*: London: Mechanical Engineering Publications.

Karaolis, N.M. 1989. The design of fibre-reinforced composite blades for passive and active wind turbine rotor aerodynamic control. PhD Thesis, Reading: The University.

Lo, K.H., McCusker, J.J. & W.G. Gottenberg 1986. Composite leaf spring for tank trailer suspensions. *41st Annual Conference, 27-31 January 1986*. The Society of the Plastics Industry: 12B/1-12B/6.

Powell, P.C. 1994. *Engineering with fibre-polymer laminates*. London: Chapman & Hall.

Pagano, N.J. (ed.) 1989. Interlaminar response of composite materials. London: Elsevier.

Pasquier, J.C. 1985. Calcul d'un arbre de transmission d'un petit vehicule automobile. In: *Matériaux et structures composites. 5eme Ecole d'hiver CODEMAC, 31 Janvier-6 Février*: Piau-Engaly.